# Forest product conversion factors

Published by
FOOD AND AGRICULTURE ORGANIZATION OF THE UNITED NATIONS
and
INTERNATIONAL TROPICAL TIMBER ORGANIZATION
and
UNITED NATIONS ECONOMIC COMMISSION FOR EUROPE

Rome, 2020

Required citation:
FAO, ITTO and United Nations. 2020. *Forest product conversion factors*. Rome. https://doi.org/10.4060/ca7952en



ISBN 978-92-5-132247-5

Designed at United Nations, Geneva – 1915745 (E) – December 2019 – 1 – ECE/TIM/NONE/2019/4/iPub

*Cover image: © Shutterstock_ID:1521053633/Juan Enrique del Barrio*

# CONTENTS

# LIST OF TABLES



# LIST OF FIGURES

# FOREWORD

The forest sector has long used conversion factors as a tool for analyzing forests and forest product manufacturing facilities. Almost every aspect of forecasting and analysis in the forest sector involves the use of conversion factors. This includes: converting from one unit of measure to another, benchmarking the efficiency of manufacturing facilities, silvicultural growth models, biomass calculations, estimates of forest carbon and timber-sale appraisals; to name just a few.

The Food and Agriculture Organization of the United Nations (FAO) recognized the need for a good understanding of conversion factors and measurement units in 1946 at the FAO Conference in Copenhagen, which was in the process of establishing a statistical database on forest products. Following this, FAO convened two conferences on forest statistics in 1947: in February in Washington, DC, United States of America, and in April 1947 in Rome, Italy. A key component of these meetings was the work of the Subcommittee on Units of Measurement, the assignment of which was explained as follows: "Because of the great diversity between units in common use for the measurement of forest products, and because the statistical programs will be world-wide in scope, it was necessary to prepare lists of converting factors". Conversion factors have since been a regular part of the work of FAO, the International Tropical Timber Organization (ITTO) and the United Nations Economic Commission for Europe (UNECE). FAO, ITTO and the UNECE regularly use conversion factors for their reporting and analysis of trade and production data on forest products. Accurate conversion factors are also critical for accurate data, in particular when conducting periodic assessments of the outlook for the forest sector.

The last study on conversion factors was published in 2010, although this covered only Europe, the Commonwealth of Independent States and North America. It was timely, therefore, for FAO, ITTO and UNECE to update existing factors, obtain a wider range of factors and improve geographic representation.

This publication builds on the 2010 study mentioned above by broadening the geographical coverage to the global level, updating factors and adding some factors that were not included in the past. Differences in measurement standards and factors have also been identified. Further investigation and cooperation will still be needed to improve the harmonization of data and factors. This has a bearing not only on conversion factors but also on the comparability of nationally reported forest product statistics.

The publication provides analysts, practitioners and private enterprises with the most up-to-date set of available forest product conversion factors and a better understanding of the units used in the manufacture, trade and reporting of wood-based forest products.

We express our appreciation to the national experts who provided their inputs and to the secretariats of FAO, ITTO and UNECE for this timely publication.

**Hiroto MITSUGI**
Assistant Director-General,
Forestry Department of the Food
and Agriculture Organization of
the United Nations

**Gerhard DIETERLE**
Executive Director,
International Tropical
Timber Organization

**Olga ALGAYEROVA**
Under-Secretary-General
of the United Nations,
Executive Secretary of the United Nations
Economic Commission for Europe

# ACKNOWLEDGEMENTS

FAO, ITTO and the UNECE thank the following people who assisted with this project:

**Mandy Allpass**, Crickmay & Associates (Pty) Ltd
**Alcinda Amaro**, AIMMP – Association of Industries of Wood Furniture, Portugal
**Jamal Balfas**, Forest Products Research and Development Center, Indonesia
**Gary Baylous**, Pacific Rim Log Scaling Bureau, United States of America
**Lars Bjorklund**, Swedish Timber Measurement Council
**Leonardo Boragno**, Ministerio de Ganadería Agricultura y Pesca, Uruguay
**Jorge Pedro Flores Marker**, Comisión Nacional Forestal, Mexico
**Peder Gjerdrum**, Norwegian Forest and Landscape Institute
**Roger Godsmark**, Forestry South Africa
**Schalk Grobbelaar**, York Timbers, South Africa
**Janina Gysling**, Instituto Forestal, Chile
**Hu Yanjie**, Research Institute of Forestry Policy and Information, Chinese Academy of Forestry
**Peter Ince**, United States Department of Agriculture, Forest Service, United States of America
**Romain Jacques**, Natural Resources Canada
**Iain Kerr**, Paper Manufacturers Association of South Africa (PAMSA)
**Ruslan Kozak**, Ukrainian National Forestry University
**Andrius Kuliesis**, State Forest Service, Lithuania
**Aleksey Kuritsin**, Lesexpert LLC, Russian Federation
**Yury Lakhtikov**, Russian Association of Pulp and Paper Organizations and Enterprises, RAO "Bumprom", Russian Federation
**Graça Louro**, Institute for Nature Conservation and Forests, Portugal
**Ludmila Lungu**, National Bureau of Statistics, Moldova
**Volodymyr Maevskyy**, Ukrainian National Forestry University
**Eleine Juliana Malek**, Timber, Tobacco and Kenaf Industries, Development Division, Ministry of Primary Industries, Malaysia
**Udo Mantau**, University of Hamburg, Germany
**Humberto Mesquita**, Brazilian Forest Service
**Anna Mohase**, Guyana Forestry Commission
**New Zealand Ministry for Primary Industries**
**Eoin O'Driscoll**, Council for Forest Research and Development, Ireland
**Jan Oldenburger**, Probos Foundation, Netherlands
**Mitja Piskur**, Slovenian Forestry Institute
**Susan Phelps**, Natural Resources Canada
**Juan Picos Martin**, Monte Industria, Spain
**Ewa Ratajczak**, PhD, Wood Technology Institute, Poland
**Matai Rewiechand**, Foundation for Forest Management and Production Control, Suriname
**Hiroyuki Saito**, Wood Products Trade Office, Forestry Agency, Japan
**Cristina Santos**, Institute for Nature Conservation and Forests, Portugal
**Peter Schwarzbauer**, University of Natural Resources and Life Sciences, Austria
**Roman Shchupakivskyy**, Ukrainian National Forestry University
**James Singh**, Guyana Forestry Commission
**Roy Southey**, Sawmilling South Africa
**Henry Spelter**, United States Department of Agriculture, Forest Service, United States of America
**Kjell Suadicani**, University of Copenhagen, Denmark
**Roman Svitok**, National Forest Centre, Slovakia

**Michal Synek**, Forest Management Institute, Czechia
**Pedro Teixera**, Centro Pinus, Portugal
**Erhabor Theophilus**, Forestry Research Institute of Nigeria
**Alain Thivolle-Cazat**, Institut Technique Forêt Cellulose Bois-Construction Ameublement, France
**Stein Tomter**, Norwegian Institute of Bioeconomy Research
**Joberto Veloso de Freitas**, Brazilian Forest Service
**Erkki Verkasalo**, Natural Resources Institute. Finland
**Darius Vizlenskas**, State Forest Service, Lithuania
**Dinko Vusić**, University of Zagreb, Faculty of Forestry, Croatia
**Sheila Ward**, Forestry Commission, United Kingdom of Great Britain and Northern Ireland
**Holger Weimar**, Federal Research Institute for Rural Areas, Forestry and Fisheries (Thünen Institute), Germany.

The following staff at FAO, ITTO and the UNECE supported the production of this publication:

**Iana Arkhipova**
**Stéphane Bothua**
**Jean-Christophe Claudon**
**Matt Fonseca**
**Arvydas Lebedys**
**Alex McCusker**
**Florian Steierer**

Matt Fonseca managed the project. Alastair Sarre edited the text. Many thanks to all the people listed above for their contributions of time and expertise. People with relevant information on forest product conversion factors that might be used to improve future revisions of this publication are invited to contact FAO, ITTO or the UNECE at the following addresses:

**Forestry Policy and Resources Division, Forestry Department**
**Food and Agriculture Organization of the United Nations (FAO)**

Viale delle Terme di Caracalla, 00153 Rome, Italy
Phone +39 (06) 570-53641
FPS@fao.org

**International Tropical Timber Organization (ITTO)**
**International Organizations Center**

5th Floor, Pacifico Yokohama, 1-1-1 Minato-Mirai
Nishi-ku, Yokohama, 220-0012 Japan
Phone +8145 223 1110
itto-stats@itto.int

**UNECE/FAO Forestry and Timber Section**
**Forests, Land and Housing Division**

Palais des Nations
CH – 1211 Geneva 10
Switzerland
Phone +41-22-917 1379
info.ece-faoforests@unece.org

# ACRONYMS AND ABBREVIATIONS

| | |
|---|---|
| **BC** | British Columbia |
| **bf** | board foot/feet |
| **C** | Celsius |
| **cm** | centimetre |
| **$CO_2$** | carbon dioxide |
| **FAO** | Food and Agriculture Organization of the United Nations |
| **ft** | foot/feet |
| **GJ** | gigajoule(s) |
| **g** | gram(s) |
| **GOST** | Government Standard of the Russian Federation |
| **ITTO** | International Tropical Timber Organization |
| **kg** | kilogram(s) |
| **m** | metre(s) |
| **$m^2$** | square metre |
| **$m^3$** | cubic metre |
| **$m^3p$** | cubic metre product |
| **$m^3p$ bulk** | cubic metre product (including void space) |
| **$m^3p$ solid** | cubic metre product (excluding void space normally included in volume) |
| **$m^3rw$** | cubic metre roundwood |
| **$m^3sw$** | cubic metre of solid wood |
| **mbf** | 1 000 board feet |
| **mcd** | moisture content dry basis |
| **mcw** | moisture content wet basis |
| **MDF** | medium-density fibreboard |
| **mm** | millimetre(s) |
| **mt** | tonne(s) |
| **NWLRAG** | Northwest Log Rules Advisory Group |
| **ob** | overbark |
| **odmt** | oven-dry tonne |
| **OSB** | oriented strandboard |
| **swe** | solid wood equivalent |
| **ub** | underbark |
| **UNECE** | United Nations Economic Commission for Europe |
| **USDA** | United States Department of Agriculture |
| **USFS** | United States Forest Service |

# EXPLANATORY NOTES

The term "softwood" is used synonymously with "coniferous". "Hardwood" is used synonymously with "non-coniferous" and "broadleaved". "Lumber" is used synonymously with "sawnwood". References to "tonnes" in this text represent the unit of 1 000 kilograms (kg). The use of "ton" in this publication represents the imperial unit of 2 000 pounds (907 kg) unless otherwise specified. "Billion" refers to 1 000 million ($10^9$). The term "oven dry" in this text is used in relation to the weight of a product in a completely dry state: e.g. 1 oven-dry tonne of wood fibre = 1 000 kg of wood fibre containing no moisture at all. Basic density is defined as the ratio between the oven-dry weight of wood and its green volume.

# CHAPTER 1

# 1. Introduction

This publication is the result of a collaborative effort between the Food and Agriculture Organization of the United Nations (FAO), the International Tropical Timber Organization (ITTO) and the United Nations Economic Commission for Europe (UNECE). It builds on and supersedes the factors contained in *UNECE/FAO (2010a)* and provides factors from all the world's major timber-producing regions.

The term "forest product conversion factors" is used to cover a broad spectrum of ratios used in the wood-based forest, manufacturing and energy sectors. For the purposes of this publication, "conversion factor" is defined as using a known figure to determine or estimate an unknown figure via a ratio. Often, these ratios are exact, for example in converting cubic feet to cubic metres (there are 35.315 cubic feet in a cubic metre). Annex table 1 provides a listing of certain exact conversion factors (equivalents) relevant to forest products.

The ratios may also be inexact – offering, rather, a good average. For example, 1 m$^3$ (underbark volume) of freshly felled Norway spruce sawlogs might have an average weight of 860 kg, of which 80 kg is bark and 780 kg is wood (with both bark and wood containing a certain amount of moisture), but this might vary as a result of, for example, wood density, moisture content and the presence or lack of bark.

In other instances, conversion factors may have little meaning unless some of the parameters of the numerators and denominators of the ratios are known. For example, 1 m$^3$ of logs with an average small-end diameter of 15 cm might make 0.41 m$^3$ of sawnwood, and 1 m$^3$ of logs with an average small-end diameter of 60 cm might make 0.63 m$^3$ (i.e. about 50 percent more), given the same level of processing efficiency in a sawmill. That is not to say, however, that a single factor cannot be used to convert roundwood to sawnwood – it can be, if using an accurate factor, when looking at a large population in the aggregate and when approximate figures are sufficient.

Related to forest product conversion factors is the use of the "material balance" (see annex figure 1). The sawnwood example above could lead to the incorrect assumption that only 41 percent of the wood fibre in a sawlog with a small-end diameter of 15 cm and 63 percent of a sawlog with a small-end diameter of 60 cm could be obtained from the log. In fact, almost 100 percent of the wood in each of these logs may be utilized because the remaining non-sawnwood volume – in the form of various wood residues – has many potential uses. For example, 1 m$^3$ of sawlogs with a small-end diameter of 15 cm could have a material balance of 41 percent sawnwood, 43 percent chips (raw material for paper, panels and wood energy, etc.), 9 percent sawdust (for making energy pellets, particle board and medium-density fibreboard (MDF), etc.) and 7 percent shavings (particle board, MDF, animal bedding and wood energy, etc.), with the various components summing to 100 percent of the wood volume. Although not part of the material balance (because the log volume comprises the underbark volume), one might also apply a conversion factor to estimate that 80 kg of bark (with moisture) is potentially available per 1 m$^3$ of roundwood (measured underbark) for energy or other uses. Note that material balances are used in manufacturing plants and at the sectoral level, and they can be constructed to account for the cascading uses of wood raw material in a country, subregion or region (Mantau, 2008).

Logs and their subsequent products have a predisposition towards inexact conversion factors because of the wide range of shapes and forms, the variability of physical properties (e.g. density, moisture content and shrinkage), and other natural variables that affect conversion factors, such as species, size, defects and provenance. Wood fibre is hygroscopic: its volume and weight change when it is dried in a kiln or exposed to the atmosphere.

Measurement procedures are subject to many external biases. For example, rounding conventions vary – with some countries determining roundwood volume based on truncated diameters and lengths and others using unbiased rounding logic. Finally, differences may occur because of product-manufacturing efficiency and utilization practices.

## 1.1 General uses of conversion factors

The forest sector has long used conversion factors as a tool for analysing forests and forest product manufacturing facilities. Almost every aspect of forecasting and analysis in

the forest sector involves the use of conversion factors in some way, including silvicultural growth models, biomass calculations, estimates of forest carbon and timber-sale appraisals.

For example, a sawmill is conducting a timber-sale appraisal to determine a bid price. The stand volume may be reported in cubic metres overbark, but the purchaser may need to convert this volume to underbark volumes, weight or board feet[1] to match its traditionally used units of measure. To determine the value of the timber, the purchaser will need to know the cost of getting the timber from the stump to the mill site; therefore, weight-to-volume ratios are likely to be important for determining weight-based transport costs. The volume of primary product recovered will need to be estimated using conversion factors (e.g. 2 m$^3$ of roundwood may produce 1 m$^3$ of sawnwood). A material balance will be used to assess the quantity and thus value of the residual products. Finally, ratios may be used to estimate the quantity of unmeasured products from the timber sale, such as bark and logging residue (i.e. top wood, limbs and foliage), which may be used for energy generation or other purposes.

Conversion factors have been used to indicate illegally logged roundwood in a supply chain – if the volume of roundwood removals is less than the apparent consumption (as determined using conversion factors), the disparity could comprise illegally harvested logs. Some organizations, including the Convention on International Trade in Endangered Species of Wild Fauna and Flora (CITES), have applied conversion factors to manufactured product volumes to estimate the harvest volume of endangered species such as bigleaf mahogany (CITES, 2008).

Conversion factors for converting the input of raw materials to the output of finished or semi-finished products are often used to benchmark the efficiency of a manufacturing facility. Climate policy analysts may use conversion factors to determine the volume of carbon sequestered in forests. Outlook studies on long-term wood availability use conversion factors to predict the volume of raw materials that will be needed to match forecasted future wood demand.

## 1.2 The use of conversion factors by FAO, ITTO and UNECE

A major objective in collecting and publishing information on conversion factors at the international level is to estimate national and international wood requirements and balances. Such information is also useful in the preparation of national and international studies involving wood balances.

Conversion factors are used in data collection, as follows:

1. **Where a country has explicitly provided data in non-standard units**. Typically, this involves a conversion from square metres to cubic metres or from tonnes to cubic metres.
2. **Where it is suspected that data are incorrect**. Often, the unit value (dividing the value of trade of a product by its volume of trade) indicates that an element is incorrect. For example, data on plywood and fibreboard are occasionally reported in square metres rather than cubic metres, without this being

1 A board foot is ostensibly the equivalent volume of a board that is 1 inch thick x 1 foot wide x 1 foot long (i.e. 0.00236 m$^3$).

explicitly stated. Here, a standard unit value (e.g. an average thickness) is applied to convert area to volume.

3. **When converting data from other sources to FAO/ ITTO/UNECE standards** (e.g. UN Comtrade[2] data reported in kilograms). Official trade statistics usually indicate trade volumes by weight (usually kilograms, although other units are also used). Thus, when extracting data from UN Comtrade, for example, weight needs to be converted from tonnes to cubic metres using conversion factors.
4. **To calculate national and international wood balances**. For example, conversion factors are used to check whether reported data from a country seem reasonable (e.g. does the net apparent consumption of wood raw materials balance with the production of wood products when wood raw materials are converted to forest products using a reasonable assumption of mill conversion factors?).
5. **To estimate future potential wood requirements under different scenarios**, including in UNECE/FAO forest sector outlook studies.

FAO, ITTO and the UNECE recognize the importance of clearly defining the units used in harmonization efforts. Such definitions are needed to obtain accurate conversion factors and also for comparisons between countries.

## 1.3 Methods

A questionnaire was sent to national correspondents to collect national conversion factors; where necessary and feasible, literature searches were conducted for countries to augment information. The following forest products were included in the questionnaire:

1. Roundwood
2. Wood particles
3. Sawnwood
4. Veneer and plywood
5. Panels made of wood particles
6. Wood pulp and paper
7. Round and split wood products
8. Energy wood products and properties

These eight forest product categories, and the relevant ratios for each included in the questionnaire, were determined primarily from recommendations made in 2008 by the Task Force on Forest Product Conversion Factors (UNECE/FAO, 2010a), which, with minor revisions, remained relevant in this publication.

Conversion factors are listed by region and country, as follows:

| Region | Country | |
|---|---|---|
| AFRICA | Nigeria | South Africa |
| ASIA | China | Japan |
| | Indonesia | Malaysia |
| EUROPE | Austria | Poland |
| | Croatia | Portugal |
| | Czechia | Russian Federation |
| | Denmark | Slovakia |
| | Finland | Slovenia |
| | France | Spain |
| | Germany | Sweden |
| | Ireland | Switzerland |
| | Lithuania | Ukraine |
| | Moldova | United Kingdom |
| | Netherlands | |
| | Norway | |
| LATIN AMERICA | Brazil | Mexico |
| | Chile | Suriname |
| | Guyana | Uruguay |
| NORTH AMERICA | Canada | United States of America |
| OCEANIA | New Zealand | |

©Shutterstock

2 The UN Commodity Trade Statistics Database is available at http://comtrade.un.org.

# CHAPTER 2

# 2. Roundwood

The conversion factors for roundwood requested in the questionnaire related to two areas:

1. physical properties; and
2. the national method of measuring roundwood as it relates to the "true volume".

Note that the true volume of roundwood for the purposes of this publication and for UNECE/FAO statistics on roundwood volume always means underbark volume. Conversion factors for roundwood are often used to convert from one unit of measure to another (e.g. from weight to volume). Conversion factors within the same units are also quite common: for example, a cubic metre of roundwood measured by the national standard in one country may differ if measured according to the national standard of another country.

Most countries were able to submit data on conversion factors relating to the general physical properties of roundwood. Note that species-specific data was not asked for; data fields were limited to coniferous and non-coniferous, with separate fields for saw and veneer logs and pulp and energy logs.

Data availability on conversion factors for harmonizing national standards of roundwood measurement to true volume was insufficient to include in this publication. Nevertheless, it is clear that there can be substantial differences in the standards used for determining log volume between countries and even within countries.

## 2.1 Volumetric measurement

Assessing the volume of roundwood is typically referred to as log scaling. In general, log scales attempt to predict the displaced volume of the log (i.e. cubic log scale), or, as in the United States of America and some regions of Canada, the output of sawnwood (e.g. board foot log rule).

A third category of log scales uses cubic (Brereton log scale) or quasi-cubic volume (Hoppus log scale) and may apply a standard ratio of board feet per cubic volumetric unit (see section 2.1.3).

### 2.1.1 Cubic volume

Cubic formulas include the Smalian, Huber, Newton, centroid and two-end conic, each of which has its strengths and weaknesses depending on log dimensions and stem form. All these formulas will give similar results most of the time when measurement conventions are applied uniformly and on logs with typical parameters. Figure 2.1 provides an example of volume calculation using unbiased rounding logic and the Smalian and Huber formulas.

Of much greater concern for harmonizing log volume and conversion factors based on national volumes is the common practice of truncating (rounding down) diameters and lengths. For example, the length of a log may be recorded as 10.0 m when its actual length is 10.3 m

FIGURE 2.1 TRUE LOG VOLUME CALCULATION

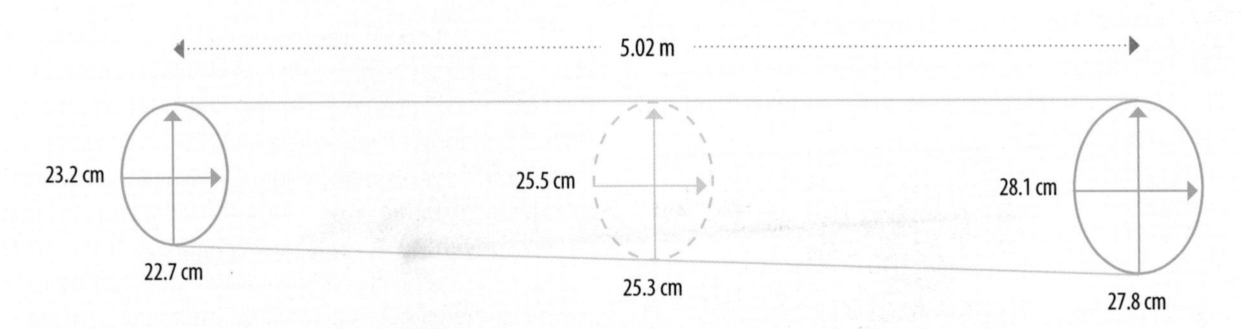


**Smalian formula**: $(((23.2+22.7)/2)^2+((28.1+27.8)/2)^2) \times 5.02 \times 0.00003927 = 2.58\ m^3$

**Huber formula**: $((25.5+25.3)/2)^2 \times 5.02 \times 0.00007854 = 2.54\ m^3$

**Source:** UNECE/FAO, 2010a.

(thus, there is 30 cm of unmeasured "trim allowance") and the diameter of a log end may be recorded as 27.0 cm when its actual diameter is 27.9 cm. The purpose of truncation should not be construed as a purposeful effort to understate volume; generally, it is done to make mathematical calculations easier and to enable the manufacture of products at least as long as the recorded log length.

Other practices also create discrepancies between log scales: for example, some systems reduce log volumes to account for defects and others use a value-reducing mechanism (such as log grade). Additionally, national and subregional log-scaling standards may: treat logs as cylinders using the small-end diameter of the log; assume that the log form is a cylinder using the diameter that exists in the middle of the log length; or use assumed taper rates to establish diameters other than the small-end. Some national standards include bark in the roundwood volume.

All these differences can lead to differences in estimates of roundwood volume. Figure 2.2 shows the saw/veneer log volume for the log in Figure 2.1, as calculated using 13 log scaling standards; it shows the considerable variation that can arise by the use of different methods. Moreover, logs with different dimensions and characteristics may produce different relative results, depending on individual characteristics: for example, truncating a diameter of 13.9 cm to 13.0 cm will have a more significant impact on the estimated volume (i.e. 12.5 percent) than truncating a diameter of 77.9 cm to 77.0 cm (2.3 percent).

Although not included in Figure 2.2, the log shown in Figure 2.1 would have a volume of:

- 30 board feet (bf) if measured in coastal Alaska, Oregon and Washington, United States of America (the Scribner long log rule);
- 40 bf if measured in the western United States of America, except as noted above (Scribner short log rule);
- 25 bf if measured in the southeast of the United States of America (Doyle log rule);
- 50 bf if measured in the northeast of the United States of America (International ¼ inch log rule);
- 45 bf if measured in Ontario, Canada (Ontario log rule);
- 48 bf if measured in New Brunswick or Nova Scotia, Canada (New Brunswick log rule); and
- 54 bf if measured in Newfoundland, Canada (Newfoundland log rule).

FIGURE 2.2 LOG VOLUME COMPARISON BETWEEN NATIONAL LOG SCALES FOR THE LOG DEPICTED IN FIGURE 2.1

(m³ and difference in percent from .258 m³)

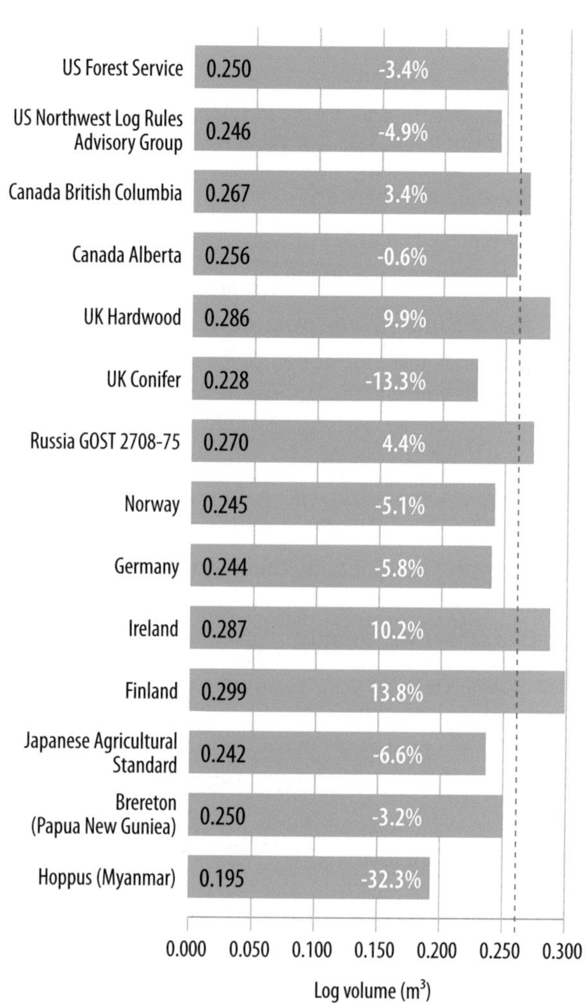


**Notes:** Calculated using the log dimensions in Figure 2.1 and applying the procedures for national roundwood measurement standards in the countries and organizations listed. GOST 2708-75 is the standard for domestically consumed roundwood in the Russian Federation; UK Conifer = top-diameter method; UK Hardwood = mid-diameter method.

**Source:** See references in subsection on roundwood measurement standards in section 10.

In this publication, conversion factors using roundwood input are assumed to be based on true volume. It is known that some countries (e.g. Finland, Ireland, Sweden, the United Kingdom of Great Britain and Northern Ireland and the United States of America) adjust their national standard roundwood measurement volumes to reflect true volume for reporting. This is an area in which further study could improve the accuracy, comparability and harmonization of what is often the starting point (the numerator) of many of the forest product conversion factors used.

### 2.1.2 Board foot log rules

Some subnational regions in the United States of America and Canada still use board foot log rules (Figure 2.3). Most of these rules date from the nineteenth century and attempt to predict the quantity of sawnwood (or "lumber"), as measured in board feet, that could be milled from a given log. In most cases, modern mills produce far more lumber (in some cases more than double the volume) to that predicted by the rules (known as overrun); occasionally, less lumber is produced than predicted by a given log rule (underrun). Log rules are based on either diagrams or formulas, with assumptions made on saw kerf and slab loss. Except for the International ¼" rule, log segments are considered to be cylindrical (i.e. have no taper).

A study in the United States of America used a population of 175 logs to model the conversion factors of various board foot rules in North America. It found, for the Scribner rules (of which there are three subregional variants) alone, that bf per m$^3$ ranged, on average, from as little as 108 (i.e. 9.26 m$^3$ per 1,000 bf[3]) for logs with a small-end diameter of 4.5 to 7.49 inches to as much as 246 (4.06 m$^3$ per mbf) for logs with a small-end diameter above 15.5 inches (Fonseca, 2005). Figure 2.4 shows the conversion (in m$^3$ per mbf) using the British Columbia standard (BC Firmwood method).[4] Because these rules do not correlate consistently with cubic volume across all diameter classes, it is necessary to view the ratios as they relate to various diameter groupings.

In the past, many sources of published conversion factors (including those of UNECE/FAO) used a universal 4.53 m$^3$ per mbf ratio (0.221 mbf per m$^3$) to convert board feet to

3 Note that "mbf" is an abbreviation for "1,000 board feet", and this abbreviation is used hereafter in this publicaton.

4 The BC Firmwood log scale was used as the index because it has unbiased rounding logic and involves the uniform application of formulas and measurements.

FIGURE 2.3 BOARD FOOT LOG RULES: GEOGRAPHIC DISTRIBUTION

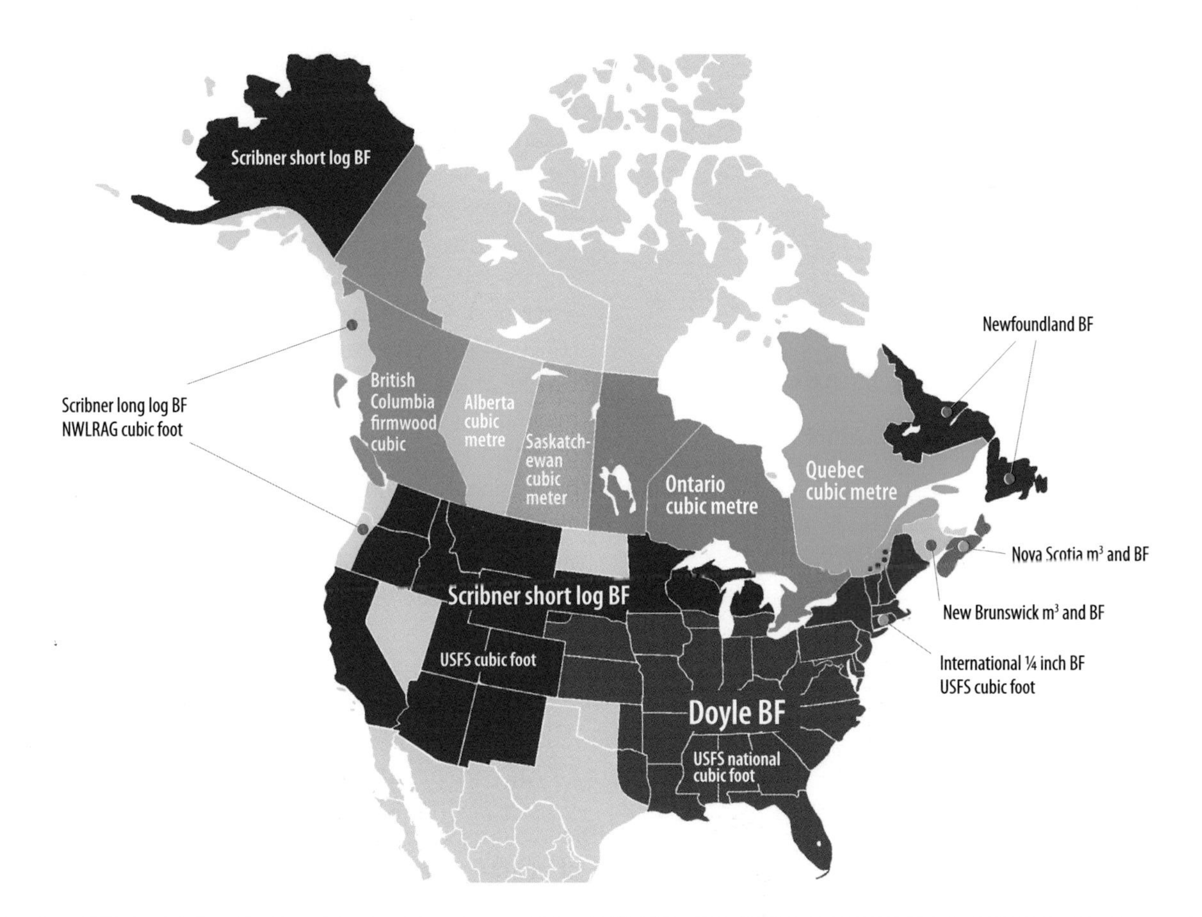


**Notes:** The states of Alaska, California, Oregon and Washington use a revised volume table, which gives slightly different volumes for several diameter and length combinations.

**Source:** UNECE/FAO, 2010b. Map conforms to United Nations *The World* map, 3-Mar-2020.

FIGURE 2.4 ESTIMATED LOG VOLUME USING VARIOUS LOG RULE METHODS AND SMALL-END DIAMETER CLASSES TO CONVERT FROM 1,000 BOARD FEET TO CUBIC METRES (m³ BC Firmwood per mbf)

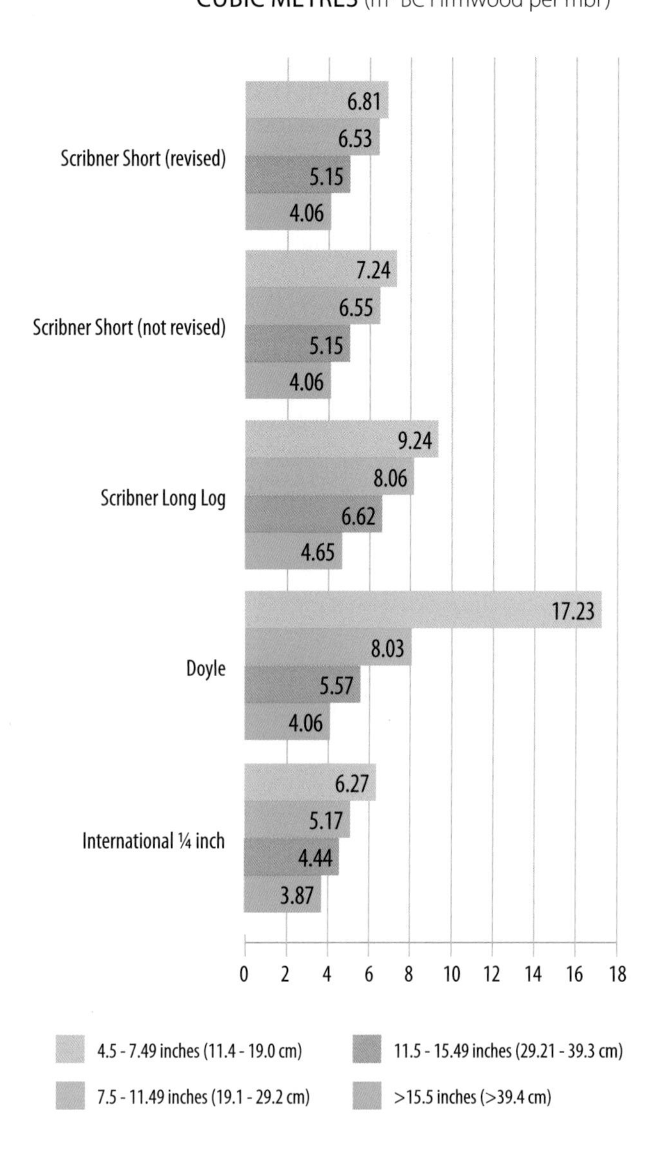


**Notes:** Cubic metres (net), as scaled by the BC Firmwood log scale, indexed in units of 1 000 board feet (mbf; net), (e.g. 6.81 m³ BC Firmwood = 1 mbf net scale). This analysis is based on a control group of 175 logs. Scribner Short (revised) = revised rule used in California and interior Oregon and Washington, USA; Scribner Short (not revised) = non-revised rule used outside the Pacific Coast states, USA; Scribner Long Log used only in coastal Alaska, Oregon and Washington, USA; Doyle used primarily in the mid-west and southeast of the USA; International ¼ inch used primarily in New England, USA, and eastern Canada.

**Source:** Fonseca, 2005.

cubic metres. This ratio dates from the 1940s (FAO, 1947), when the timber harvest on the west coast of North America involved very-large-diameter trees, and it is responsible for countless erroneous assumptions made by researchers and practitioners who have attempted to use it. In reality, conversion factors vary widely based on the scaling standards used (both for m³ and bf standards), log attributes (e.g. diameter, length, defects and taper), and how the log is manufactured. There is substantial variation between the various board foot log rules and in the logs they are applied to; an approach stratified by geographic area and log-scaling method, therefore, appears to be most appropriate. Table 2.1 shows the historical development of cubic-content-to-mbf conversion factors in North America.

Spelter (2002) found that the conversion factor for the Scribner long log mbf (a variation of the Scribner log rule used in the US Pacific Northwest) to cubic metres had changed from about 4.69 m³ per mbf in 1970 to 6.71 m³ per mbf in 1998. More recent studies conducted by the Pacific Rim Log Scaling Bureau on 3.7 million dual-scaled logs in 2010–2015 showed conversion factors in the range of 6.68–6.92 m³ per mbf (6.83 average) (Pacific Rim Log Scaling Bureau, personal communication, 2016), which supports the findings of Spelter (2002). In the Washington state interior (Scribner short log method), the ratio increased from 4.42 m³ per mbf in 1970 to 5.93 m³ per mbf in 1998 (Spelter, 2002).

In summary, there is no universal figure for converting logs measured in board feet to cubic volume (and vice versa). The ratio will vary significantly depending on the board foot rule used, and there will also be substantial variation due to characteristics such as log diameter, length, taper and defect. Exacerbating the difficulty of using a ratio to convert between board foot and cubic log rules is the way in which logs are manufactured from tree stems to minimize or maximize the impacts of log characteristics on board foot log scaling volumes (either increasing or decreasing the volume). For rough estimates across a wide area and variety of forest types, Table 2.1 contains what is probably the most up-to-date, comprehensive set of conversion factors for converting from board feet to cubic metres in the geographic regions listed.

### 2.1.3 Brereton and Hoppus log scales

#### *2.1.3.1 Brereton*

The Brereton system of measurement is essentially a cubic log volume formula and therefore not dissimilar to many of the other systems of cubic measure. The Brereton log scale, however, uses its own methodology for determining

TABLE 2.1 HISTORICAL DEVELOPMENTS OF $M^3$ TO MBF CONVERSION FACTORS BY LOG SCALE AND REGION

| *Source* | *Year* | *Log scale* | *Region* | *$m^3$/mbf* | *Sample size* |
|---|---|---|---|---|---|
| FAO | 1947 | Not specified | USA | 4.53 | Unknown |
| Spelter | 1970 | Long log Scribner | Western Washington | 4.69 | Calculated* |
| Spelter | 1998 | Long log Scribner | Western Washington | 6.76 | Calculated* |
| Pacific Rim Log Scaling Bureau | 2010 | Long log Scribner | Western Oregon, Washington | 6.92 | 363 378 $m^3$ |
| Pacific Rim Log Scaling Bureau | 2011 | Long log Scribner | Western Oregon, Washington | 6.84 | 568 566 $m^3$ |
| Pacific Rim Log Scaling Bureau | 2012 | Long log Scribner | Western Oregon, Washington | 6.86 | 793 435 $m^3$ |
| Pacific Rim Log Scaling Bureau | 2013 | Long log Scribner | Western Oregon, Washington | 6.76 | 937 188 $m^3$ |
| Pacific Rim Log Scaling Bureau | 2014 | Long log Scribner | Western Oregon, Washington | 6.85 | 667 425 $m^3$ |
| Pacific Rim Log Scaling Bureau | 2015 | Long log Scribner | Western Oregon, Washington | 6.68 | 151 386 $m^3$ |
| Spelter | 1970 | Short log Scribner | Eastern Washington | 4.42 | Calculated* |
| Spelter | 1998 | Short log Scribner | Eastern Washington | 5.93 | Calculated* |
| Fonseca | 2001-2003 | Short log Scribner | Montana | 6.74 | 4.1 million $m^3$ |
| Brandt *et al.* | 2006 | Short log Scribner | Idaho | 5.99 | Calculated* |
| Mclever *et al.* | 2012 | Short log Scribner | California | 5.25 | Calculated* |
| Pacific Rim Log Scaling | 2013 | Short log Scribner | Eastern Oregon | 5.28 | 33 105 $m^3$ |
| Pacific Rim Log Scaling | 2014 | Short log Scribner | Eastern Oregon | 5.47 | 19 075 $m^3$ |
| Nova Scotia Natural Resources | 2001 | New Brunswick log rule | Nova Scotia | 5.66 | Unknown |

**Notes:** British Columbia Firmwood $m^3$ is the standard used in this table (except for Nova Scotia, which uses its own cubic log scale). Other methods of measuring cubic volume would likely yield different results. *Calculated ratios are determined using sawnwood production and material balance to calculate the input of cubic log volume. The figures for Montana include a substantial quantity (7.3 percent by gross volume) of "salvage logs" cut from standing dead trees, primarily beetle-killed lodgepole pine. This list only includes conversion factors for converting logs manufactured for scale in board feet because board foot rules are sensitive to log length, minimum top size, taper and log defects (and thus are manufactured with log volume in mind). Logs manufactured using measurement in cubic metres will likely yield different values if converted to board feet. All examples are in the USA except for Nova Scotia (Canada).

**Sources:** FAO, 1947; Spelter, 2002; Pacific Rim Log Scaling Bureau, personal communication, 2016; Fonseca, personal communication, 2006; Brandt *et al.*, 2012; Mclever *et al.*, 2015; Nova Scotia Department of Natural Resources, 2001.

log cross-sections; moreover, in some regions, it is still reflected in board feet, which is simply the volume in cubic feet × 12. The Brereton log scale is one of the most commonly used methods for tropical hardwoods in Africa, Oceania, South America and Asia (South East Asia Lumber Producers Association, 1981; Freese, 1973). Several national and international standards use Brereton as the basis of log volume calculation, with the methodologies differing slightly in their approaches to assessing lengths and diameters and whether reflected in board feet, cubic feet or cubic metres.

### *2.1.3.2 Hoppus*

Hoppus, also called the "quarter girth formula" (or the "Francon system" when measured with and reflected in metric units), became popular as a method for assessing roundwood volume in the early 1700s and is still used in the tropical regions of Asia, Africa and South America and occasionally in Australia, New Zealand and the United Kingdom of Great Britain and Northern Ireland. The Hoppus formula is:

**(middle circumference of log in inches ÷ 4)$^2$ × length in feet ÷ 144 = Hoppus cubic feet**

Hoppus cubic feet may be further converted to "Hoppus superficial feet" (i.e. Hoppus cubic feet × 12 – ostensibly converting cubic feet to board foot); or the "Hoppus ton" (i.e. Hoppus volume in cubic feet ÷ 50). Both the Hoppus superficial foot and the Hoppus ton are misnomers, because the former in no way predicts output in board feet and the latter is likely to weigh significantly more than

a ton (or tonne). Note that because the Hoppus formula essentially uses one-fourth of the log circumference as a *de facto* cross-dimension of a square, it will always reflect only 78.54 percent of the true cubic volume (assuming consistent measurement methodology) (Freese, 1973).

### 2.1.4 Stacked measure

In estimating the volume of stacked roundwood, it is often assumed that about two-thirds (66.7 percent) of the displaced volume is wood, 11.5 percent is bark, and 21.6 percent is void (Ontario Ministry of Natural Resources, 2000). In fact, wood volume may constitute less than 50 percent of the volume or more than 80 percent, depending on a range of factors. A guide for adjusting for such factors can be found in *Estimation of the Solid Volume Percentage* (VMF Nord, 1999) and also in Table 2.2, which allows the estimation of the solid wood ratio based on visually assessed attributes. Two units of measure are associated with stacked measure volume:

1. *stere*, which is 1 $m^3$ of stacked wood (including void space and bark); and
2. *cord*, which is 128 cubic ft of stacked wood (including void space and bark) and equivalent to 3.62 steres.

TABLE 2.2 SWEDISH NATIONAL BOARD OF FORESTRY STACKED MEASURE GUIDELINES FOR PULP LOGS

**Starting value 60%**

| Average log diameter under bark | |
|---|---|
| <8.99 cm | -4% |
| 9–9.99 cm | -3% |
| 10–10.99 cm | -2% |
| 11–11.99 cm | -1% |
| 12–12.99 cm | 0% |
| 13–13.99 cm | +1% |
| 14–14.99 cm | +2% |
| 15–15.99 cm | +3% |
| 16–16.99 cm | +4% |
| 17–17.99 cm | +5% |
| 18–19.99 cm | +6% |
| 20–22.99 cm | +7% |
| 23–26.99 cm | +8% |
| >27 cm | +9% |

| Quality of stacking | |
|---|---|
| Very good | 0% |
| Good | -1% |
| Fair | -2% |
| Bad | -3 to -4% |
| Very bad | -5% to -7% |

| Trimming of knots and buttress | |
|---|---|
| Very well trimmed, few buttresses | 0% |
| A few knot stumps and buttresses | -1% |
| Many knot stumps and buttresses | -2% |
| Many large knot stumps, clusters and buttresses | -3% to +4% |
| Very bad trimming | -5% to +7% |

| Crookedness | |
|---|---|
| Straight softwood, straight aspen | 0 to -1% |
| Crooked softwood, straight deciduous wood | -2% to +3% |
| Very crooked softwood, crooked deciduous wood | -4% to +5% |
| Very crooked deciduous wood | -6% to +7% |
| Extremely crooked deciduous wood | -8% to +12% |

| Bark volume | |
|---|---|
| Debarked wood | +7% |
| More than ⅔ thin bark | +2% |
| More than ⅓ thin bark | +1% |
| Normal softwood (conifer) | 0% |
| Softwood with thick bark, normal deciduous bark | -1% |
| More than 50% non-coniferous (deciduous) wood with thick bark | -2% |

**Notes:** A more recent and detailed version is available. The 1999 version was used in this publication for conciseness, however, the reader is encouraged to refer to "SDC's instruction for timber measurement: measurement of roundwood stacks" (SDC, 2014).

**Source:** VMF Nord, 1999.

## 2.2 Weight and physical properties

Roundwood is bought and sold by weight in many areas of the world. The presence of drive-on weight scales on many transportation routes and at processing plants and national ports of entry means that weight data are often more readily available and less expensive to obtain than individually measuring all logs for volume. Weight generally correlates well with volume, and the relationship is usually established in conjunction with sample volume measurements. A few factors are key in determining the weight of a given roundwood volume, and vice versa: wood density, moisture content, and bark.

### 2.2.1 Wood density

Wood largely comprises cell walls and void spaces. Wood cell walls all have about the same basic density, regardless of species – approximately 1 560 kg per m$^3$ (Van Vuuren *et al.*, 1978). What varies is the ratio of cell walls to cell cavities.

Wood density is typically measured as either basic density (i.e. the weight of oven-dry fibre per cubic metre, where "oven dry" means devoid of moisture) or specific gravity, which is an index of the relationship of oven-dry fibre to the same volume of water (water weighs 1 000 kg per m$^3$). For example, 1 m$^3$ of wood (volume measured when green) without water may weigh 400 kg; it therefore has a basic density of 400 kg per m$^3$ and a specific gravity of 0.40. Globally, the basic density of wood (green volume and oven-dry weight) varies from about 290 kg per m$^3$ to 540 kg per m$^3$ for coniferous wood and from about 160 kg per m$^3$ to more than 1 000 kg per m$^3$ for non-coniferous wood, with most species falling in the range of 320–720 kg per m$^3$ (United States Forest Service, 1999).

Care should be taken when using published averages to calculate the basic density or specific gravity of a species to note whether the volume was established in the green state (i.e. before shrinkage) or in the dry state (after shrinkage). Wood volume will shrink volumetrically by about 10 percent for most coniferous species and by about 15 percent for most hardwood species when taken from green to fully dry (although the difference could be as low as 7 percent to more than 20 percent). The exact amount of shrinkage varies between species and even between wood samples from the same tree. In general, basic density based on green volume (the standard for this publication unless otherwise noted) has the advantage of being applicable to standing tree volume and roundwood without needing to know or estimate volumetric shrinkage. When using published volumetric shrinkage rates, it is generally accepted that shrinkage in wood is initiated at moisture contents below 30 percent moisture content "dry basis" (mcd), and the amount of shrinkage is related linearly to the moisture content as it relates to the 30 percent figure. For example, a species that averages 10 percent volumetric shrinkage when taken from a green to a dry state that is dried to 15 percent mcd will lose about 5 percent of its volume when dried and thus will be densified (because what was once 1 m$^3$ of wood would now be only 0.95 m$^3$).

### 2.2.2 Moisture content

Freshly cut wood contains large amounts of water, both in the cell cavities (free water) and within the cell walls themselves (bound water). Normally, wood moisture content is measured in terms of the weight of the moisture relative to the weight of the dry wood fibre. For example, if a sample of wood weighs 812 kg per m$^3$ in its green state and 400 kg per m$^3$ in the oven-dry state, it is said to have 103 percent mcd. In this example, the wood has 400 kg of wood and 412 kg of moisture, and dividing the weight of the moisture by the weight of the dry wood gives the moisture content. Note that the mcd of freshly cut wood can vary from 30 percent to more than 200 percent (United States Forest Service, 1999).

Moisture content can also be reflected in the ratio of moisture weight to the total weight of the wood fibre plus moisture content, referred to as moisture content "wet basis" (mcw). This is typically used for measuring the moisture content of wood particles (Chapter 2), pulp and paper (Chapter 6) and wood for energy (Chapter 8). It is not usually used for solid wood because the numerator is made up partly by the denominator.

The moisture content of roundwood is dynamic and can decline quickly once a tree is felled, especially in hot,

dry weather. Section 2.3 lists weight-to-volume ratios for freshly felled logs (green weight with bark/green m$^3$ (wood only)), but these ratios may not be representative in regions where logs are stored for long periods at roadside before transport and weighing. Note also that the moisture content of the wood and bark of many species has distinct seasonal variations, with significantly higher moisture content in winter and early spring than in summer and early fall (Thivolle-Cazat, 2008).

Finally, in many species, particularly conifers, heartwood often has less moisture than sapwood. Thus, older (generally larger) trees often have a lower weight-to-volume ratio than that of younger (generally smaller) trees by virtue of the age-related increase in the ratio of heartwood to sapwood. Related to this, many species have a higher ratio of bark when young and small than when old and large, thereby magnifying this difference.

### 2.2.3 Bark and other unmeasured volume

Bark, like wood, typically contains large amounts of water when fresh. In general, the bark of most species has weight-to-volume characteristics similar to the wood of the same species. The bark of both coniferous and non-coniferous species ranges from less than 5 percent of the total overbark volume (and weight) to as high as 30 percent. Based on the questionnaire results shown in section 2.3, bark constitutes 11–12 percent, on average, of the overbark volume. Roundwood weight is normally reported with bark present; however, roundwood volume for global reporting purposes, including in the majority of national roundwood measurement standards, is reported for wood only, meaning that bark increases the weight-to-volume ratio of roundwood. Bark is an important source of forest-based energy, and it also has other uses, such as decorative ground cover and soil treatment (mulch). Knowing the ratio of bark to roundwood volume is useful for understanding weight ratios but also potentially for the energy and other products that can be produced from bark. It is important to note, however, that some bark volume is almost always lost during handling from forest to mill, so the potential volume (based on subtracting underbark from overbark volume, as measured in standing inventories) is seldom available and highly volatile depending on log-handling practices and season. Another point to consider when assessing bark volume, especially when using overbark-minus-underbark volume, is that the bark of many species has fissures and voids, which can lead to incorrect estimates of bark volume unless taken into account.

Defects in roundwood, such as unsound fibre (decay), fractures, splits and crooked portions, have weight but often no volume when the scaling method calls for a deduction, thus increasing the ratio of weight to volume for roundwood.

### 2.2.4 Carbon content and sequestered carbon dioxide

It is generally accepted that, although there are variations in the carbon content of wood, the average figure is close to 50 percent of the dry weight of wood and bark and that this figure is accurate enough for carbon accounting purposes (Matthews, 1993). There is potential for confusion between the carbon and carbon dioxide ($CO_2$) content of wood, which are not the same. The ratio of $CO_2$ to carbon weight is 3.67: thus, 1 tonne of dry wood sequesters about 0.5 tonnes of carbon and 1.83 tonnes of $CO_2$. If using carbon or $CO_2$ ratios that have been developed for volumetric units rather than dry weight, care needs to be taken to account for the effect of shrinkage on the dry weight of a given volume, which would densify the wood and thus increase the quantity of contained carbon per unit volume (see section 2.2.1).

## 2.3 Summary of country data on roundwood

### 2.3.1 Africa, Asia, Latin America, North America and Oceania

| PRODUCTS | unit in/ unit out | Africa | | | Asia | | | | | | Latin America | | | | | | | | North America | Oceania |
|---|---|---|---|---|---|---|---|---|---|---|---|---|---|---|---|---|---|---|---|---|
| | | Nigeria | South Africa | **Median/ average** | China | Indonesia | Japan | Malaysia | **Median** | **Average** | Brazil | Chile | Guyana | México | Suriname | Uruguay | **Median** | **Average** | United States of America | New Zealand |
| **SAW/VENEER LOGS** | | | | | | | | | | | | | | | | | | | | |
| **Conifer** | share % | | 100 | | 62 | 2 | 99 | | | | | 97 | | 70 | | 14 | | | 73 | 100 |
| Green weight with bark/green m³ (wood only) | kg/m³ | | 940 | **940** | 940 | 740 | | | **840** | **840** | | 872 | | 900 | | 882 | **882** | **885** | 1 000 | 1 000 |
| Wood basic density (dry weight of wood/green m³ wood only) | kg/m³ | | 430 | **430** | 420 | 420 | | | **420** | **420** | 385 | 405 | | 450 | | 405 | **405** | **411** | 455 | 433 |
| Volume ratio wood/bark plus wood | ub/ob | | 0.95 | **0.95** | 0.88 | 0.85 | | | **0.87** | **0.87** | 0.85 | 0.92 | | 0.90 | | 0.88 | **0.89** | **0.89** | 0.88 | 0.85 |
| **Non-conifer** | share % | 100 | | | 38 | 98 | 1 | 100 | | | | 3 | 100 | 30 | 100 | 86 | | | 27 | |
| Green weight with bark/green m³ wood only | kg/m³ | 1 184 | | **1 184** | 1 060 | 1 130 | | 1 156 | **1 130** | **1 115** | | 1 368 | 1 180 | 1 120 | 1 152 | 910 | **1 152** | **1 146** | 1 086 | |
| Wood basic density (dry weight of wood/green m³ wood only) | kg/m³ | 675 | | **675** | 620 | 540 | | 680 | **620** | **613** | 500 | 625 | 843 | 700 | 783 | 489 | **663** | **657** | 527 | |
| Volume ratio wood/bark plus wood | ub/ob | 0.91 | | **0.91** | 0.85 | 0.87 | | 0.88 | **1** | **1** | 0.85 | 0.92 | 0.79 | 0.85 | 0.92 | 0.90 | **0.88** | **0.87** | 0.88 | |
| **PULP LOGS** | | | | | | | | | | | | | | | | | | | | |
| **Conifer** | share % | | | | 20 | 3 | 53 | | | | | 57 | | 70 | | | | | 60 | 100 |
| Green weight with bark/green m³ wood only | kg/m³ | | | | 820 | 720 | | | **770** | **770** | | 872 | | 850 | | | **861** | **861** | 855 | 1 000 |
| Wood basic density (dry weight of wood/green m³ wood only) | kg/m³ | | 450 | **450** | 410 | 410 | | | **410** | **410** | 410 | 405 | | 400 | | | **405** | **405** | 444 | 433 |
| Volume ratio wood/bark plus wood | ub/ob | | | | 0.88 | 0.87 | | | **0.88** | **0.88** | 0.85 | 0.92 | | 0.90 | | | **0.90** | **0.89** | 0.89 | 0.85 |
| **Non-conifer** | share % | | | | 80 | 97 | 47 | | | | | 43 | | 30 | | 100 | | | 40 | |
| Green weight with bark/green m³ wood only | kg/m³ | | | | 890 | 1 090 | | | **990** | **990** | | 1 368 | | 960 | | 900 | **960** | **1 076** | 893 | |
| Wood basic density (dry weight of wood/green m³ wood only) | kg/m³ | | 550 | **550** | 460 | 525 | | | **493** | **493** | 460 | 625 | | 600 | | 449 | **530** | **534** | 471 | |
| Volume ratio wood/bark plus wood | ub/ob | | | | 0.90 | 0.85 | | | **0.88** | **0.88** | 0.85 | 0.92 | | 0.90 | | 0.90 | **0.90** | **0.89** | 0.89 | |

**Notes:** The share of conifer to non-conifer reflects reported industrial roundwood and is included so that a weighted average for both can be calculated; ob = overbark; ub = underbark.

**Source:** FAO/ITTO/UNECE Forest product conversion factors questionnaire, 2018.

## 2.3.2 Europe

| PRODUCTS | Unit in/ unit out | Europe | | | | | | | | | | | | | | | | | | | | | | |
|---|---|---|---|---|---|---|---|---|---|---|---|---|---|---|---|---|---|---|---|---|---|---|---|---|
| | | Croatia | Czechia | Denmark | Finland | France | Germany | Ireland | Lithuania | Moldova | Netherlands | Norway | Poland | Portugal | Russian Federation | Slovakia | Slovenia | Spain | Sweden | Switzerland | Ukraine | UK | **Median** | **Average** |
| **SAW/VENEER LOGS** | | | | | | | | | | | | | | | | | | | | | | | | |
| **Conifer** | share (%) | 20 | 95 | 91 | 96 | 73 | 93 | 100 | 68 | 1 | 77 | 100 | 83 | 98 | 70 | 78 | 85 | 72 | 99 | 90 | 83 | 99 | | |
| Green weight with bark/green m³ (wood only) | kg/m³ | 979 | 940 | 870 | 930 | 970 | 779 | | 919 | 720 | 780 | 820 | 800 | 989 | 815 | 940 | 875 | 881 | 950 | 900 | 870 | 1 018 | **891** | **887** |
| Wood basic density (dry weight of wood/green m³ wood only) | kg/m³ | 371 | 434 | 390 | 410 | 420 | 389 | 360 | 420 | 460 | 440 | 400 | 480 | 368 | 517 | | 405 | 436 | 415 | 440 | 430 | | **420** | **420** |
| Volume ratio wood/bark plus wood | ub/ob | 0.89 | 0.90 | 0.90 | 0.89 | 0.87 | 0.89 | 0.89 | 0.90 | 0.88 | 0.82 | 0.90 | 0.88 | 0.8 | 0.90 | 0.90 | 0.89 | 0.87 | 0.90 | | 0.87 | 0.89 | **0.89** | **0.88** |
| **Non-conifer** | share (%) | 80 | 5 | 9 | 4 | 27 | 7 | | 32 | 99 | 23 | | 17 | 2 | 30 | 22 | 15 | 28 | 1 | 10 | 17 | 1 | | |
| Green weight with bark/green m³ wood only | kg/m³ | 1 003 | 1 150 | 1 020 | 1 050 | 1 100 | 1 125 | | 899 | 1 015 | 940 | 900 | 911 | 1 000 | 870 | 1 180 | 1 050 | 902 | 1 050 | 1 100 | 1 050 | 1 143 | **1 035** | **1 023** |
| Wood basic density (dry weight of wood/green m³ wood only) | kg/m³ | 553 | 680 | 560 | 510 | 560 | 563 | | 474 | 610 | 420 | 500 | 614 | | 640 | | 570 | 554 | 550 | 625 | 550 | | **560** | **561** |
| Volume ratio wood/bark plus wood | ub/ob | 0.91 | 0.87 | 0.90 | 0.88 | 0.88 | 0.91 | | 0.90 | 0.87 | 0.85 | 0.85 | 0.88 | 0.8 | 0.89 | 0.89 | 0.94 | 0.88 | 0.88 | | 0.85 | 0.88 | **0.88** | **0.88** |
| **PULP LOGS** | | | | | | | | | | | | | | | | | | | | | | | | |
| **Conifer** | share (%) | 30 | 93 | 100 | 76 | 56 | 49 | 100 | 59 | | 57 | 98 | 75 | 8 | 50 | 50 | 50 | 38 | 91 | 60 | | 100 | | |
| Green weight with bark/green m³ wood only | kg/m³ | 983 | 940 | 870 | 850 | 950 | 779 | | 931 | | 780 | 820 | 800 | 1057 | 740 | 940 | 900 | 882 | 920 | 900 | | 1018 | **900** | **892** |
| Wood basic density (dry weight of wood/green m³ wood only) | kg/m³ | 382 | 434 | 390 | 400 | 407 | 389 | 360 | 420 | | 440 | 400 | | 500 | 450 | | 405 | 436 | 400 | 444 | | | **406** | **416** |
| Volume ratio wood/bark plus wood | ub/ob | 0.89 | 0.90 | 0.90 | 0.87 | 0.83 | 0.89 | 0.89 | 0.89 | | 0.82 | 0.90 | 0.88 | 0.74 | 0.90 | 0.90 | 0.87 | 0.84 | 0.90 | | | 0.89 | **0.89** | **0.87** |
| **Non-conifer** | share (%) | 70 | 7 | 0 | 24 | 44 | 51 | | 41 | | 43 | 2 | 25 | 92 | 50 | 50 | 50 | 62 | 9 | 40 | | 0 | | |
| Green weight with bark/green m³ wood only | kg/m³ | 997 | 1 150 | 1 020 | 950 | 1 050 | 1 125 | | 929 | | 940 | 900 | 911 | 976 | 800 | 1 180 | 1 100 | 1 155 | 970 | 1 100 | | 1 143 | **1 009** | **1 022** |
| Wood basic density (dry weight of wood/green m³ wood only) | kg/m³ | 564 | 680 | 560 | 490 | 550 | 563 | | 474 | | 420 | 500 | 614 | 516 | 570 | | 570 | 620 | 500 | 625 | | | **562** | **551** |
| Volume ratio wood/bark plus wood | ub/ob | 0.91 | 0.87 | 0.90 | 0.86 | 0.85 | 0.91 | | 0.89 | | 0.85 | 0.85 | 0.87 | 0.82 | 0.89 | 0.89 | 0.91 | 0.85 | 0.88 | | | 0.88 | **0.88** | **0.88** |

**Notes:** The share of conifer to non-conifer reflects reported industrial roundwood and is included so that a weighted average for both can be calculated; ob = overbark; ub = underbark.

**Source:** FAO/ITTO/UNECE Forest product conversion factors questionnaire, 2018.

©Shutterstock

# CHAPTER 3

# 3. Wood particles

Wood particles (chips, sawdust, flakes and shavings) can be measured by volume and weight (both in the dry state and "as delivered"). All wood-particle products start out as solid wood from logs of varying density, are broken down into somewhat irregularly shaped particles, and often contain varied amounts of moisture and void space between particles.

Conversion factors for wood particles are determined by wood density, moisture content and compaction. In general, large enterprises that produce and use wood particles obtained from many sources use oven-dry weight for measuring wood particles. Enterprises that work with wood particles obtained from fewer and more homogeneous sources might favour the use of volume or weight as delivered. Ultimately, the yield of most manufacturing processes using wood particles as a raw material is driven by the quantity of fibre, excluding moisture and void.

Another driver of the yield of wood particles is classification. Often, users of wood particles require classification by size aimed at ensuring the suitability of particles for the product they are manufacturing (e.g. some pulping processes and oriented strandboard – OSB). A sawmill or wood-chipping operation may need to remove small particles (fines) from "on specification" chips, which otherwise would result in a lower yield of chips and a higher yield of small particles; these small particles may be classified and marketed as sawdust.

## 3.1 Volumetric measurement

The procedure for establishing the volume of particles is straightforward. Volumes contained in truckloads, ship hulls and bins can easily be calculated. Even huge, irregularly shaped stockpiles at manufacturing facilities can be measured with surveying equipment to establish volume.

Solid wood equivalent is more complicated, however, because of the variation in void space. Wood particles in containers or piles will settle over time and the heavier the particles (due to density or moisture content) and the greater the depth (thus increasing weight), the more compaction will occur.

Based on data obtained from the questionnaire, the median volume of wood particles per cubic metre of solid wood is 2.80 $m^3$ for chips and 2.87 $m^3$ for sawdust (meaning that 1 $m^3$ of solid wood will produce chips that displace 2.80 $m^3$, including void space). The following are listed in *Conversion Factors for the Pacific Northwest Forest Industry* (Hartman *et al.*, 1981) as ratios of $m^3$ loose to $m^3$ solid for wood particles:

- pulp chips (compacted) 2.50
- pulp chips (uncompacted) 2.86
- sawdust 2.50
- planer shavings 4.00

## 3.2 Weight

The weight of wood particles is generally given "as delivered" or as oven-dry weight (all moisture removed, often called "bone dry" in North America). The "as delivered" weight may vary substantially due to differences in moisture content. For example, chips produced in a sawmill may have 50 percent mcw (i.e. 50 percent of the weight is water and 50 percent is dry fibre), but a veneer plant using logs identical to those used in the sawmill might produce chips in which only 6 percent (mcw) of the

©UNECE

weight is moisture. In this example, the sawmill chips were green and the veneer chips were produced from dried veneer. Based on the questionnaire results, the median ratio (all countries) of delivered tonnes of wood chips is about 2.0 for chips (delivered tonne to oven-dry tonne) and sawdust (indicating 50 percent dry fibre, 50 percent mcw); and 1.16 for shavings (indicating 86 percent dry fibre and 14 percent mcw).

The normal procedure for establishing the oven-dry weight of wood particles is via a sampling system, as per the following example:

- The net weight of a truckload of chips is 32 200 kg.
- A sample of "as delivered" chips is taken, weighing 922 g.
- The sample is placed in a vented oven at approximately 103 °C for 24 hours, until the weight stabilizes at 497 g (devoid of moisture).
- The oven-dry weight of the sample is divided by the "as delivered" weight and this ratio is multiplied by the "as delivered" net weight of the truckload of chips.
- i.e. (497 ÷ 922) × 32 200 = 17 356 kg or 17.356 oven-dry tonnes.

## 3.3 Summary of country data on wood particles

### 3.3.1 Africa, Asia, Latin America, North America and Oceania

| | Unit in/ unit out | *Africa* | | | *Asia* | | | | | *Latin America* | | | | | *North America* | *Oceania* |
|---|---|---|---|---|---|---|---|---|---|---|---|---|---|---|---|---|
| | | Nigeria | South Africa | **Median/ average** | China | Indonesia | Malaysia | **Median** | **Average** | Chile | Mexico | Uruguay | **Median** | **Average** | United States of America | New Zealand |
| **WOOD CHIPS** | | | | | | | | | | | | | | | | |
| Green swe to oven-dry tonne | m³/odmt | | 2.44 | **2.44** | 2.3 | 2.60 | 2.40 | **2.40** | **2.43** | 2.19 | 2.22 | 2.23 | **2.22** | **2.21** | 2.31 | 2.30 |
| Average delivered tonne/odmt | mt/odmt | | 2.19 | **2.19** | 2.15 | 2.15 | 2.00 | **2.15** | **2.10** | 1.52 | | | **1.52** | **1.52** | 2.15 | |
| m³ loose to solid m³ | m³/m³ | | 3.76 | **3.76** | 3.05 | 2.80 | 2.80 | **2.80** | **2.88** | 2.33 | 2.86 | | **2.60** | **2.60** | 3.04 | |
| **SAWDUST** | | | | | | | | | | | | | | | | |
| Green swe to oven dry tonne | m³/odmt | 2.02 | 2.44 | **2.23** | 2.38 | 2.60 | 2.40 | **2.40** | **2.46** | | 2.22 | 2.23 | **2.23** | **2.23** | 2.31 | 2.30 |
| Average delivered tonne/odmt | mt/odmt | 1.83 | 2.19 | **2.01** | 2.16 | 2.15 | 2.00 | **2.15** | **2.10** | | | | | | 2.15 | |
| m³ loose to solid m³ | m³/m³ | 2.36 | 3.76 | **3.06** | 3.05 | 2.80 | 2.80 | **2.80** | **2.88** | 3.23 | 2.50 | | **2.87** | **2.87** | 3.04 | |
| **SHAVINGS** | | | | | | | | | | | | | | | | |
| Green swe to oven dry tonne | m³/odmt | | 2.44 | **2.44** | 2.34 | 2.60 | 2.30 | **2.34** | **2.41** | | 2.22 | 2.23 | **2.23** | **2.23** | 2.31 | 2.30 |
| Average delivered tonne/odmt | mt/odmt | | 1.15 | **1.15** | 1.16 | | 1.15 | **1.16** | **1.16** | | | 1.15 | **1.15** | **1.15** | 1.15 | |
| m³ loose to solid m³ | m³/m³ | | 3.76 | **3.76** | | | | | | 3.23 | | | **3.23** | **3.23** | | |

**Notes:** odmt = oven-dry tonne; swe = solid wood equivalent (assumes green volume of wood, prior to any shrinkage); loose m³ = indicates bulk volume (including the void space between wood particles).

**Source:** FAO/ITTO/UNECE Forest product conversion factors questionnaire, 2018.

### 3.3.2 Europe

| | Unit in/ unit out | *Europe* | | | | | | | | | | | | | | | | | | |
|---|---|---|---|---|---|---|---|---|---|---|---|---|---|---|---|---|---|---|---|---|
| | | Czechia | Denmark | Finland | France | Germany | Ireland | Lithuania | Netherlands | Norway | Poland | Portugal | Russian Federation | Slovakia | Slovenia | Spain | Sweden | Switzerland | **Median** | **Average** |
| **WOOD CHIPS** | | | | | | | | | | | | | | | | | | | | |
| Green swe to oven dry tonne | m³/odmt | 2.41 | | 2.40 | 2.25 | 2.49 | 2.78 | 2.29 | | 2.50 | | 2.86 | 2.40 | 2.44 | 2.20 | 2.45 | 2.30 | 2.25 | **2.41** | **2.43** |
| Average delivered tonne/odmt | mt/odmt | 2.00 | | 2.00 | 1.85 | | 2.11 | | | 1.80 | 2.22 | | | 1.98 | 1.80 | | 2.00 | | **2.00** | **1.97** |
| m³ loose to solid m³ | m³/m³ | 2.88 | 2.70 | 2.55 | 2.50 | | 2.70 | 2.78 | 2.86 | 3.30 | 2.86 | | 2.78 | 3.38 | 3.00 | | 2.70 | 2.80 | **2.80** | **2.84** |
| **SAWDUST** | | | | | | | | | | | | | | | | | | | | |
| Green swe to oven dry tonne | m³/odmt | 2.41 | | 2.40 | 2.25 | 2.49 | 2.78 | 2.29 | | 2.50 | | 3.33 | 2.29 | 2.44 | 2.34 | 2.45 | 2.30 | 2.25 | **2.41** | **2.47** |
| Average delivered tonne/odmt | mt/odmt | 2.00 | | 2.00 | 1.50 | | 2.11 | | | 1.80 | 2.22 | | | 1.98 | 1.80 | | 2.00 | | **2.00** | **1.93** |
| m³ loose to solid m³ | m³/m³ | 2.88 | | 2.55 | 3.00 | | 2.70 | 3.03 | 2.86 | 3.30 | 2.86 | | 3.03 | 3.38 | 3.00 | | 2.70 | 2.80 | **2.88** | **2.93** |
| **SHAVINGS** | | | | | | | | | | | | | | | | | | | | |
| Green swe to oven dry tonne | m³/odmt | 2.41 | | 2.40 | 2.25 | 2.49 | 2.78 | 2.29 | | 2.50 | | | | 2.44 | | 2.45 | 2.30 | 2.25 | **2.41** | **2.37** |
| Average delivered tonne/odmt | mt/odmt | | | | | | | | | | | | | | | | | | | |
| m³ loose to solid m³ | m³/m³ | | | | | | | | | | | | | | | | | | | |

**Notes:** odmt = oven-dry tonne (oven dry tonne); swe = solid wood equivalent (assumes green volume of wood, prior to any shrinkage); loose m³ = indicates bulk volume (including the void space between wood particles).

**Source:** FAO/ITTO/UNECE Forest product conversion factors questionnaire, 2018.

# CHAPTER 4

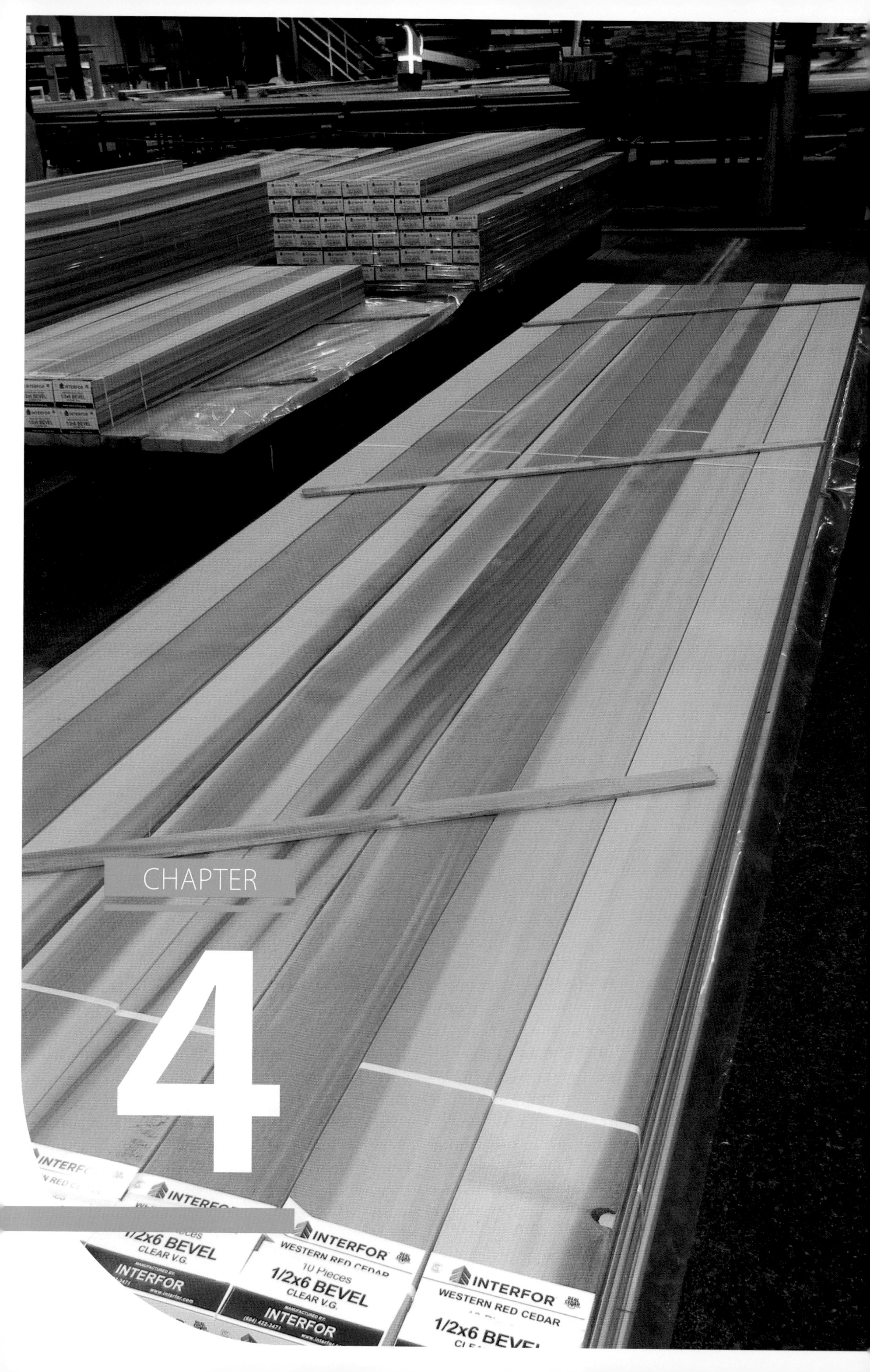

INTERFOR
WESTERN RED CEDAR
10 Pieces
1/2x6 BEVEL
CLEAR V.G.
MANUFACTURED BY:
INTERFOR
INTERFOR
WESTERN RED CEDAR
1/2x6 BEVEL

# 4. Sawnwood

Sawnwood covers a large spectrum of sawn solid wood products, ranging from semi-processed cants,[5] boules and flitches,[6] which are often still green, to fully edged, dried, trimmed and planed sawnwood. The state of manufacture (e.g. boules/flitches versus dried and planed sawnwood) has a large impact on the conversion factor: for example, 1 m³ of roundwood might produce 0.8 m³ of green boules and flitches but only half that amount (0.4 m³) of fully dried, edged, grade-trimmed and surfaced (planed) sawnwood. Thus, countries were asked to provide conversion factors for subcategories of sawnwood products in order to understand the considerable variation between national conversion factors for sawnwood.

5 A cant is a semi-processed log with at least one (generally two or four) flat faces (either sawn or chipped).

6 A flitch is sawnwood that has not yet had the edgings removed; thus, the wide face is tapered lengthwise and includes the rounded profile of the log on its edges. A boule is a log manufactured into flitches and stacked together into a unit resembling the original log.

For example, Germany reported a conversion factor for coniferous sawnwood of 1.59 m³ of roundwood per m³ of sawnwood, and the United States of America reported a factor of 2.04. On face value, this might indicate that sawmills in the United States of America are much less efficient than sawmills in Germany (because the values suggest that mills in the former require 28 percent more roundwood volume to make the same quantity of sawnwood as mills in the latter). A further analysis of the subcategories of sawnwood, however, shows that roundwood-to-sawnwood conversion factors for subproduct categories such as rough green and surfaced dry are similar for both countries. What differs is the point in the manufacturing process at which sawnwood production volume is measured. It may also be that Germany produces more rough wood (i.e. not surfaced by planing) than does the United States of America.

A significant number of correspondents reported that sawnwood production in their countries is measured in

FIGURE 4.1 EXAMPLE OF ROUNDWOOD-TO-SAWNWOOD CONVERSION FACTORS, BY STATE OF MANUFACTURE

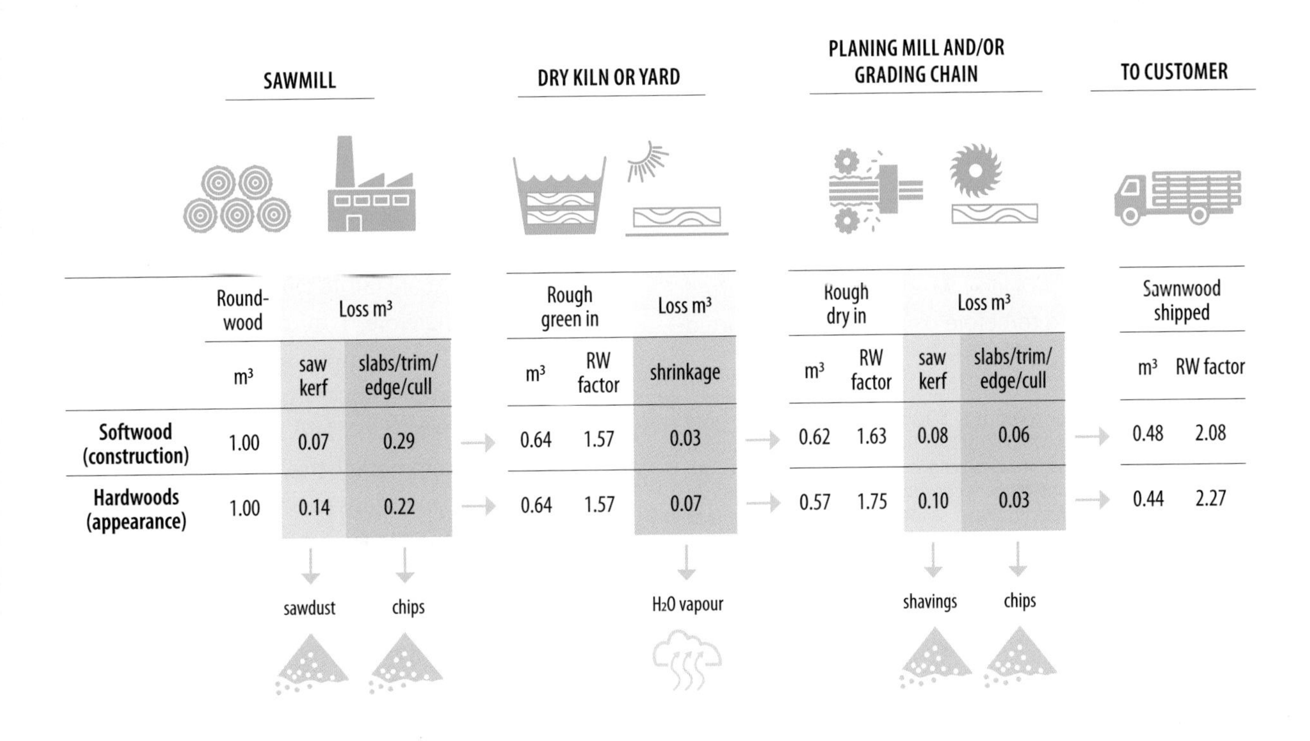


| | Roundwood | Loss m³ | | Rough green in | | Loss m³ | Rough dry in | | Loss m³ | | Sawnwood shipped | |
|---|---|---|---|---|---|---|---|---|---|---|---|---|
| | m³ | saw kerf | slabs/trim/edge/cull | m³ | RW factor | shrinkage | m³ | RW factor | saw kerf | slabs/trim/edge/cull | m³ | RW factor |
| Softwood (construction) | 1.00 | 0.07 | 0.29 | 0.64 | 1.57 | 0.03 | 0.62 | 1.63 | 0.08 | 0.06 | 0.48 | 2.08 |
| Hardwoods (appearance) | 1.00 | 0.14 | 0.22 | 0.64 | 1.57 | 0.07 | 0.57 | 1.75 | 0.10 | 0.03 | 0.44 | 2.27 |

**Source:** UNECE/FAO, 2010a.

the rough green state to avoid "double counting" volume, which may be dried, grade-trimmed or planed at separate off-site facilities. This contrasts with other countries, notably in the Nordic region and North America, where sawnwood volume is more often tallied and reported in a final state of manufacture. This can lead to a situation in which the ratio of roundwood needed to make sawnwood is reported as 1.57 (64 percent recovery) in the rough green state; 1.75 (57 percent recovery) in the rough dry state; and 2.27 (44 percent) in a fully planed and finished state (Figure 4.1).

In addition to the state of manufacture, many other drivers of efficiency can affect conversion factors for roundwood to sawnwood. The issue of volumetric measurement is discussed in section 4.1, but log quality and size have substantial impacts on conversion efficiency, as do differences in how roundwood volume is measured (see section 2.1). In addition, recovery ratios are affected by the efficiency of the milling process and the types of product made. Figure 4.2 shows the typical sawnwood product recovery and material balance for dried and planed dimension lumber[7] (sawnwood) in North America. Note the substantial improvement in sawnwood recovery as diameter increases.

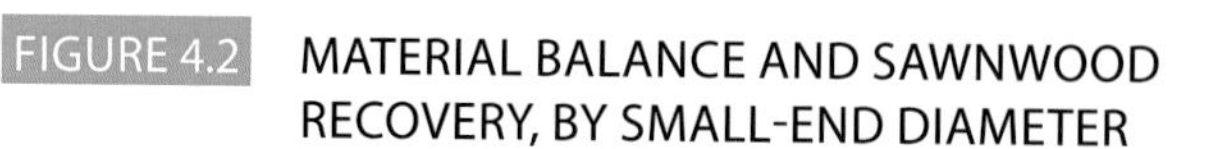


MATERIAL BALANCE AND SAWNWOOD RECOVERY, BY SMALL-END DIAMETER

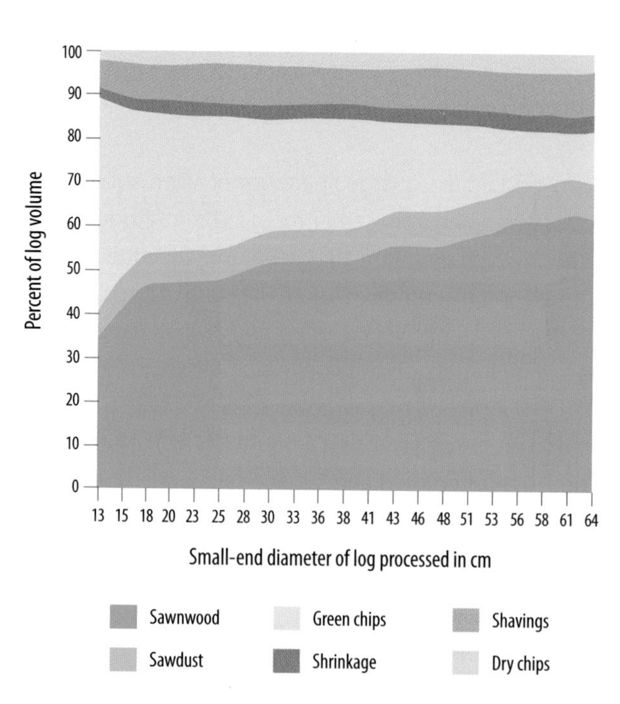


**Notes:** Based on studies at dimension lumber mills in the northwestern USA. Green chips are produced in the sawmill, and dry chips come from trimming or edging to remove defects (after the wood is dried).

**Source:** Fonseca, 2005.

## 4.1 Volumetric measurement

### 4.1.1 Cubic volume

Sawnwood is measured in cubic metres in most of the world outside North America. It is generally assumed that volumes are determined based on actual dimensions allowing for small variations (e.g. ± 2 mm). The formula is:

width in mm × thickness in mm × length in m ÷ 1 000 000 = $m^3$ sawnwood

In some areas of Europe, sawnwood is sold in the retail market based on volumes determined by nominal sizes.[8] For example, the volume of dried and planed lumber that is 45 mm × 95 mm, and which might have started out at 50 mm × 100 mm before planing, will be reported based on the latter dimensions. Where the ratio of actual-to-nominal volume can be assessed accurately, for example in North America for sawn softwood, production statistics and conversion factors are adjusted to reflect actual volume; where they are not known, however, no adjustments are made.

### 4.1.2 Board foot volume

In North America, sawnwood is usually measured in board feet, with a board foot defined as a board 1 inch (2.54 cm) thick and 1 foot square (30.48 cm × 30.48 cm), equivalent to 0.00236 $m^3$. Therefore, 1 $m^3$ of sawnwood ostensibly equals 424 bf. This measure arose when sawnwood was commonly sold in rough, green form. Today, sawn softwood is typically dried and surfaced at the sawmill before sale. To accommodate this, a set of standards allows wood of smaller dimensions to be sold on a nominal basis that ignores the shrinkage and material removed during surfacing. For example, the volume of dried and planed sawnwood may be calculated based on the nominal dimensions of 2 inches thickness × 4 inches width × 96

7 Dimension lumber (sawnwood) is a term used in North America to classify sawnwood which is 2 inches or greater in nominal thickness and is for use in applications where structural strength and serviceability are the primary concerns. It is normally produced in lengths from 8–20 feet (in two-foot increments) and in widths of 4–12 inches (in two-inch increments).

8 Nominal size is the approximate or roughcut dimensions by which a material is generally called or sold in trade but which differs from the actual dimensions due to shrinkage in the drying process and the removal of rough and uneven surfaces.

TABLE 4.1 BOARD FOOT ACTUAL-TO-NOMINAL SIZES AND VOLUMES, BY LUMBER (SAWNWOOD) PRODUCT TYPE

| | *Nominal* | | | *Green target size** | | | *Dried-planed size* | | |
|---|---|---|---|---|---|---|---|---|---|
| **PRODUCT** | **Thick (inches)** | **Width (inches)** | **bf/m³** | **Thick (inches)** | **Width (inches)** | **bf/m³** | **Thick (inches)** | **Width (inches)** | **bf/m³** |
| **SOFTWOOD BOARDS** | | | | | | | | | |
| 1×4 | 1 | 4 | 424 | 0.94 | 3.8 | 475 | 0.75 | 3.5 | 646 |
| 1×6 | 1 | 6 | 424 | 0.94 | 5.88 | 461 | 0.75 | 5.5 | 616 |
| 1×8 | 1 | 8 | 424 | 0.94 | 7.88 | 458 | 0.75 | 7.25 | 624 |
| 1×10 | 1 | 10 | 424 | 0.94 | 9.88 | 457 | 0.75 | 9.25 | 611 |
| 1×12 | 1 | 12 | 424 | 0.94 | 11.88 | 456 | 0.75 | 11.25 | 603 |
| **DIMENSION AND STUDS** | | | | | | | | | |
| 2×4 | 2 | 4 | 424 | 1.75 | 3.8 | 510 | 1.5 | 3.5 | 646 |
| 2×6 | 2 | 6 | 424 | 1.75 | 5.88 | 495 | 1.5 | 5.5 | 616 |
| 2×8 | 2 | 8 | 424 | 1.75 | 7.88 | 492 | 1.5 | 7.25 | 624 |
| 2×10 | 2 | 10 | 424 | 1.75 | 9.88 | 491 | 1.5 | 9.25 | 611 |
| 2×12 | 2 | 12 | 424 | 1.75 | 11.88 | 489 | 1.5 | 11.25 | 603 |
| **SOFTWOOD SHOP** | | | | | | | | | |
| 5/4 | 1.25 | actual | 424 | 1.417 | actual | 374 | 1.156 | actual | 458 |
| 6/4 | 1.5 | actual | 424 | 1.68 | actual | 378 | 1.406 | actual | 452 |
| 7/4 | 1.75 | actual | 424 | 1.878 | actual | 395 | 1.594 | actual | 465 |
| 8/4 | 2 | actual | 424 | 2.108 | actual | 402 | 1.813 | actual | 468 |
| 9/4 | 2.25 | actual | 424 | 2.404 | actual | 397 | 2.094 | actual | 456 |
| 10/4 | 2.5 | actual | 424 | 2.7 | actual | 392 | 2.375 | actual | 446 |
| **HARDWOOD LUMBER** | | | | | | | | | |
| 2/4 | 0.5 | actual | 424 | 0.605 | actual | 701 | 0.313 | actual | 1,356 |
| 3/4 | 0.75 | actual | 424 | 0.874 | actual | 485 | 0.563 | actual | 753 |
| 4/4 | 1 | actual | 424 | 1.142 | actual | 371 | 0.813 | actual | 522 |
| 5/4 | 1.25 | actual | 424 | 1.411 | actual | 375 | 1.063 | actual | 499 |
| 6/4 | 1.5 | actual | 424 | 1.68 | actual | 378 | 1.313 | actual | 484 |
| 7/4 | 1.75 | actual | 424 | 1.882 | actual | 394 | 1.5 | actual | 494 |
| 8/4 | 2 | actual | 424 | 2.151 | actual | 394 | 1.75 | actual | 484 |
| 9/4 | 2.25 | actual | 424 | 2.487 | actual | 384 | 2.063 | actual | 462 |
| 10/4 | 2.5 | actual | 424 | 2.688 | actual | 394 | 2.25 | actual | 471 |
| 11/4 | 2.75 | actual | 424 | 2.957 | actual | 394 | 2.5 | actual | 466 |
| 12/4 | 3 | actual | 424 | 3.226 | actual | 394 | 2.75 | actual | 462 |
| 13/4 | 3.25 | actual | 424 | 3.495 | actual | 394 | 3 | actual | 459 |
| 14/4 | 3.5 | actual | 424 | 3.763 | actual | 394 | 3.25 | actual | 456 |
| 15/4 | 3.75 | actual | 424 | 4.032 | actual | 394 | 3.5 | actual | 454 |
| 16/4 | 4 | actual | 424 | 4.301 | actual | 394 | 3.75 | actual | 452 |

**Notes:** * Green target sizes are set by the manufacturer based on shrinkage, size control, etc., and thus these are reasonable averages. Softwood "shop grades" (used as stock to extract cuttings for further manufacture, such as door and window components) and hardwood lumber – widths are based on actual, which is commonly referred to as "random width", and thickness is often reflected in ¼ inch increments – e.g. 5/4 has a nominal thickness of 1.25".

**Sources:** Western Wood Products Association, 1998; National Hardwood Lumber Association, 1994; Fonseca, 2005.

inches length (0.0126 m³), but its actual dimensions may be 1.5 inches thickness × 3.5 inches width × 92.625 inches length (0.008 m³). The formula for determining board feet is:

Nominal width in inches × nominal thickness in inches × nominal length in feet ÷ 12 = bf

Thus, a board that is nominally 2 inches × 4 inches × 8 feet would comprise 5.333 bf.

For measuring sawn softwood volume, the width, thickness and, to a smaller degree, length have nominal measurement allowances. In other words, what nominally is represented as having 1 m³ is often as little as 0.66 m³. This fact is accounted for in the FAOSTAT and UNECE/FAO timber databases using an actualization factor of 72 percent. Thus, it is assumed there are 589 bf per m³.

The standards for sawn hardwood in North America differ from those for sawn softwood and are considered to be truer to the implied 0.00236 m³ per bf (424 bf per m³); thus, no actualization factor is used. Often, sawn hardwood is sawn oversize to the nominal thickness, and widths and lengths are based on actual dimensions. Additionally, a significant component of sawn hardwood is produced in rough form.

The board feet contained in a cubic metre varies according to the state of manufacture of the sawnwood (e.g. green, dry and rough, or surfaced and dry) and the type of sawnwood manufactured (Table 4.1).

## 4.2 Weight

The relationship between the weight and volume of sawnwood varies because of basic density, moisture content and shrinkage (see sections 1.2.1–1.2.3). The relationship is useful, however, for estimating shipping weight from known volumes and volumes from known shipping weights. It is also common to estimate this ratio using the following formula (Briggs, 1994):

(basic density ÷ (1 – shrinkage)) × (1 + moisture content)

For example, the calculation of weight per volume for Scots pine sawnwood originating in a region where it has an average basic density of 400 kg per m³ (volume measured before shrinkage), and a volumetric shrinkage of about 7.5 percent for sawnwood dried to 15 percent mcd, is as follows:

$$(400 \div (1 - 0.075)) \times (1 + 0.15) = 497.3 \text{ kg/m}^3$$

## 4.3 Material balance

In completing the questionnaire, many countries submitted the material balance for the manufacture of sawnwood. The sawnwood component accounts for roughly half the roundwood volume that is input to the sawmilling process, and it is important, therefore, to understand the residual products that later become the raw materials for other wood products – such as chips and slabs, sawdust and shavings. Figure 4.3 and Figure 4.4 show the material balances for those countries that reported it.

FIGURE 4.3 MATERIAL BALANCE IN THE SAWMILLING PROCESS FOR CONIFEROUS SAWNWOOD, BY REPORTING COUNTRY (percentage)

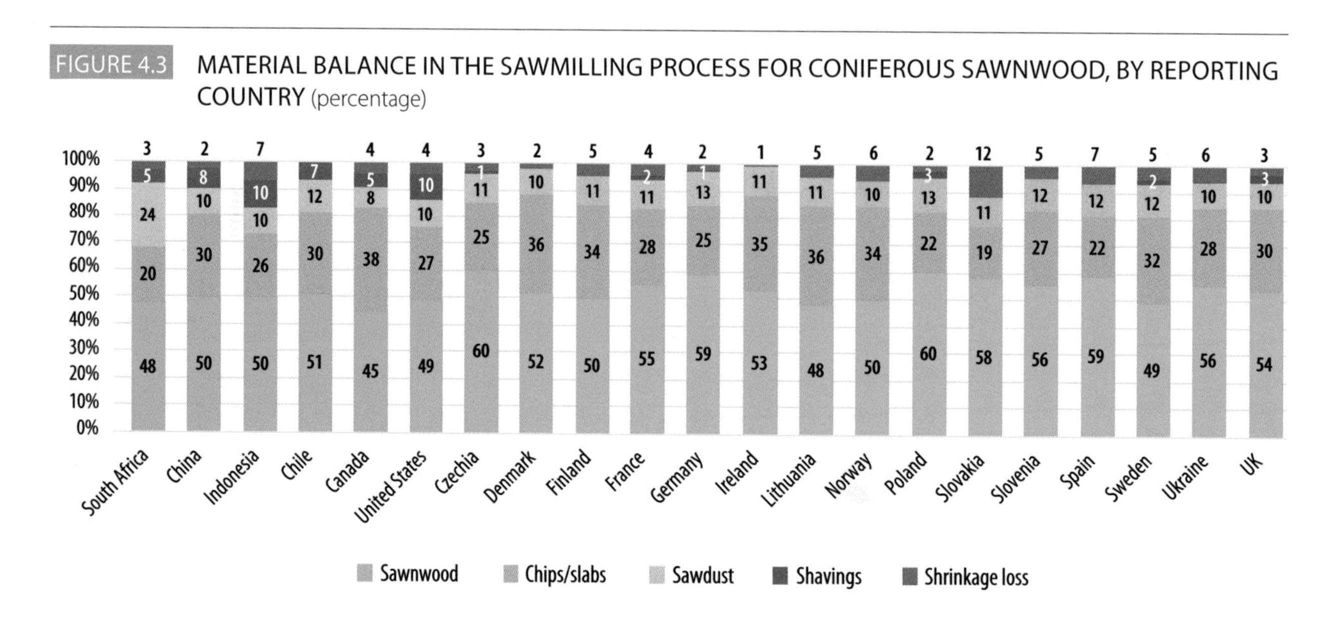


**Source** : FAO/ITTO/UNECE Forest product conversion factors questionnaire, 2018.

FIGURE 4.4 MATERIAL BALANCE IN THE SAWMILLING PROCESS FOR NON-CONIFEROUS SAWNWOOD
(percentage)

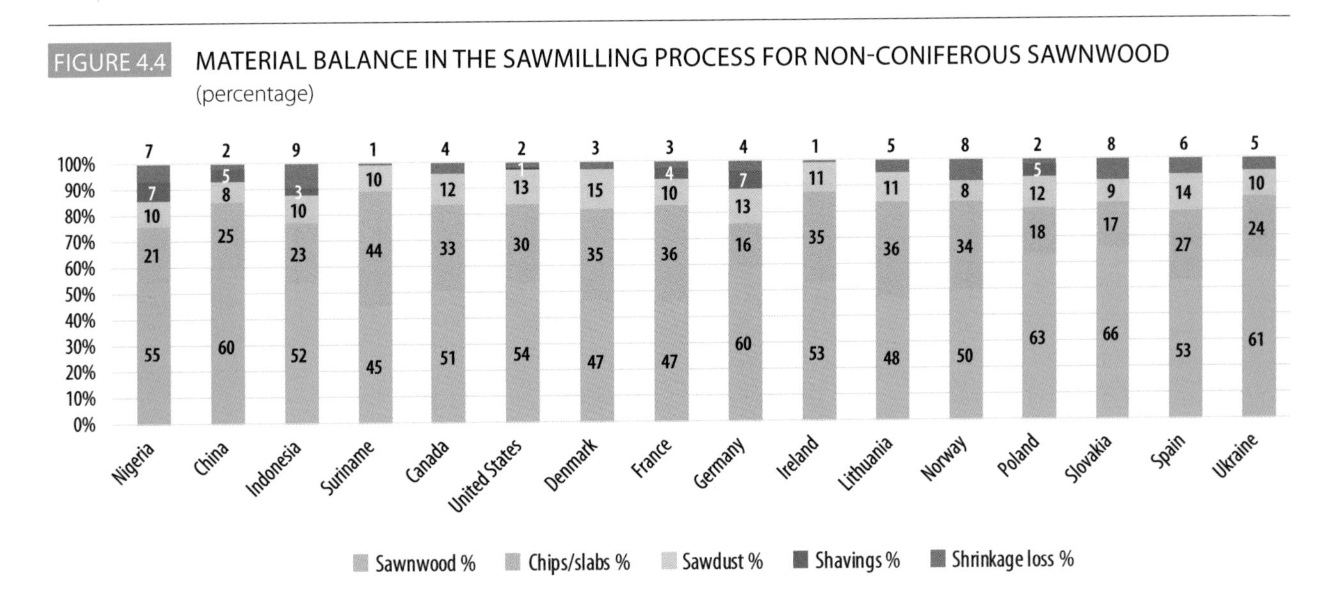


**Source:** FAO/ITTO/UNECE Forest product conversion factors questionnaire, 2018.

## 4.4 Summary of country data on sawnwood

### 4.4.1 Africa, Asia, Latin America, North America and Oceania

| | | Africa | | | Asia | | | | | | Latin America | | | | | | | North America | | | Oceania |
|---|---|---|---|---|---|---|---|---|---|---|---|---|---|---|---|---|---|---|---|---|---|
| | **Unit in/ unit out** | Nigeria | South Africa | **Median/ Average** | China | Indonesia | Japan | Malaysia | **Median** | **Average** | Brazil | Chile | Guyana | Suriname | Uruguay | **Median** | **Average** | Canada | United States of America | **Median/ Average** | New Zealand |
| **CONIFER** | Share (%) | | 100 | | 68 | 2 | 90 | | | | | 97 | | | 57 | | | 97 | 68 | | 100 |
| **Sawnwood all** | m³rw/m³p | | 2.08 | **2.08** | 2.00 | 2.00 | 1.57 | | **2.00** | **1.86** | 2.22 | | | | 2.02 | **2.12** | **2.12** | 2.22 | 2.04 | **2.13** | 1.81 |
| Sawnwood green rough | m³rw/m³p | | 1.66 | **1.66** | 1.71 | 1.65 | | | **1.68** | **1.68** | | 1.97 | | | | **1.97** | **1.97** | | 1.62 | **1.62** | |
| Sawnwood green planed | m³rw/m³p | | | | 2.05 | 2.00 | | | **2.03** | **2.03** | | 2.28 | | | | **2.28** | **2.28** | | 2.00 | **2.00** | |
| Sawnwood dry rough | m³rw/m³p | | 1.75 | **1.75** | 1.72 | 1.70 | | | **1.71** | **1.71** | | | | | | | | | 1.69 | **1.69** | |
| Sawnwood dry planed | m³rw/m³p | | 2.08 | **2.08** | 2.11 | 2.10 | | | **2.11** | **2.11** | | | | | | | | | 2.08 | **2.08** | |
| Flitches and boules (green rough) | m³rw/m³p | | | | 1.40 | 1.25 | | | **1.33** | **1.33** | | | | | | | | | | | |
| Flitches and boules (dry rough) | m³rw/m³p | | | | 1.51 | 1.50 | | | **1.51** | **1.51** | | | | | | | | | | | |
| **Material balance** | | | | | | | | | | | | | | | | | | | | | |
| Sawnwood | % | | 48 | **48** | 50 | 50 | | | **50** | **50** | 45 | 51 | | | 49 | **49** | **48** | 45 | 49 | **47** | |
| Chips/slabs | % | | 20 | **20** | 30 | 26 | | | **27** | **27** | | 30 | | | | **30** | **30** | 38 | 27 | **33** | |
| Sawdust | % | | 24 | **24** | 10 | 10 | | | **10** | **10** | | 12 | | | | **12** | **12** | 8 | 10 | **9** | |
| Shavings | % | | 5 | **5** | 8 | 10 | | | **9** | **9** | | 7 | | | | **7** | **7** | 5 | 10 | **8** | |
| Shrinkage loss | % | | 3 | **3** | 2 | 7 | | | **5** | **5** | | | | | | | | 4 | 4 | **4** | |
| Average sawnwood shipping weight | kg/m³ | | 430 | **430** | 592 | 520 | | | **556** | **556** | 420 | 508 | | | | **464** | **464** | | 581 | **581** | |
| **NON-CONIFER** | Share (%) | 100 | | | 32 | 98 | 10 | 100 | | | | 3 | 100 | 100 | 43 | | | 3 | 32 | | |
| **Sawnwood all** | m³rw/m³p | 1.82 | | **1.82** | 1.92 | 1.90 | 1.82 | 1.59 | **1.86** | **1.81** | | | | 2.29 | 1.91 | **2.20** | **2.10** | 1.96 | 1.85 | **1.91** | |
| Sawnwood green rough | m³rw/m³p | 2.15 | | **2.15** | 1.81 | 2.10 | | 1.59 | **1.81** | **1.83** | 2.22 | 1.93 | 2.20 | 2.29 | | **2.21** | **2.16** | | 1.79 | **1.79** | |
| Sawnwood green planed | m³rw/m³p | | | | | 2.00 | | | **2.00** | **2.00** | | 2.23 | | | | **2.23** | **2.23** | | | | |
| Sawnwood dry rough | m³rw/m³p | | | | 1.98 | 1.70 | | | **1.84** | **1.84** | | | | | | | | | 1.92 | **1.92** | |
| Sawnwood dry planed | m³rw/m³p | | | | 2.42 | 2.10 | | | **2.26** | **2.26** | | | | | | | | | 2.38 | **2.38** | |
| Flitches and boules (rough green) | m³rw/m³p | | | | 1.40 | 1.30 | | | **1.35** | **1.35** | | | | | | | | | | | |
| Flitches and boules (rough dry) | m³rw/m³p | | | | 1.51 | 1.50 | | | **1.51** | **1.51** | | | | | | | | | | | |
| **Material balance** | | | | | | | | | | | | | | | | | | | | | |
| Sawnwood | % | 55 | | **55** | 60 | 52 | | 63 | **60** | **58** | 45 | 52 | 40 | 45 | 52 | **45** | **47** | 51 | 54 | **53** | |
| Chips/slabs | % | 21 | | **21** | 25 | 23 | | | **24** | **24** | | | | 44 | | **44** | **44** | 33 | 30 | **32** | |
| Sawdust | % | 10 | | **10** | 8 | 10 | | | **9** | **9** | | | | 10 | | **10** | **10** | 12 | 13 | **13** | |
| Shavings | % | 7 | | **7** | 5 | 3 | | | **4** | **4** | | | | | | | | 0 | 1 | **1** | |
| Shrinkage loss | % | 7 | | **7** | 2 | 9 | | 8 | **8** | **6** | | | | | | | | 4 | 2 | **3** | |
| Average sawnwood shipping weight | kg/m³ | | | | 720 | 670 | | 800 | **720** | **730** | 650 | 858 | | | | **754** | **754** | | | | |

**Notes:** m³rw = cubic metre roundwood; m³p = cubic metre product. / **Source**: FAO/ITTO/UNECE Forest products conversion factors questionnaire, 2018.

## 4.4.2 Europe

| | Unit in/ unit out | Austria | Croatia | Czechia | Denmark | Finland | France | Germany | Ireland | Lithuania | Netherlands | Norway | Poland | Portugal | Russian Federation | Slovakia | Slovenia | Spain | Sweden | Ukraine | UK | **Median** | **Average** |
|---|---|---|---|---|---|---|---|---|---|---|---|---|---|---|---|---|---|---|---|---|---|---|---|
| **CONIFER** | Share (%) | 98 | 17 | 95 | 83 | 99 | 83 | 94 | 100 | 68 | 70 | 99 | 89 | 98 | | 73 | 87 | 73 | 99 | 74 | 98 | | |
| **Sawnwood all** | m³rw/m³p | 1.66 | 1.54 | 1.67 | 1.92 | 2.00 | 1.81 | 1.67 | 1.89 | 2.08 | 1.88 | 2.00 | 1.64 | 2.00 | | 1.72 | 1.75 | 1.69 | 2.04 | 1.78 | 1.85 | **1.81** | **1.82** |
| Sawnwood green rough | m³rw/m³p | | | | | | 1.65 | 1.66 | 1.89 | | 1.88 | 1.75 | 1.63 | | 1.85 | 1.54 | 1.75 | 1.69 | | | | **1.69** | **1.71** |
| Sawnwood green planed | m³rw/m³p | | | | | | 2.05 | | | | | | | | | | | | | | | **2.05** | **2.05** |
| Sawnwood dry rough | m³rw/m³p | | | | | 2.00 | 1.80 | | 1.99 | | | 2.00 | | | | 1.72 | 1.85 | 1.90 | 2.04 | 2.05 | 2.00 | **2.00** | **1.94** |
| Sawnwood dry planed | m³rw/m³p | | | | | | 2.40 | 1.99 | 2.13 | | | 2.50 | | | | | | | | | | **2.27** | **2.26** |
| Flitches and boules (green rough) | m³rw/m³p | | | | | | 1.18 | | | | | | | | | 1.33 | | | | | | **1.26** | **1.26** |
| Flitches and boules (dry rough) | m³rw/m³p | | | | | | 1.50 | | | | | | | | | 1.49 | | | | | | **1.50** | **1.50** |
| **Material balance** | | | | | | | | | | | | | | | | | | | | | | | |
| Sawnwood | % | 60 | | 60 | 52 | 50 | 55 | 59 | 53 | 48 | 53 | 50 | 60 | 50 | | 58 | 56 | 59 | 49 | 56 | 54 | **55** | **55** |
| Chips/slabs | % | | | 25 | 36 | 34 | 28 | 25 | 35 | 36 | | 34 | 22 | | | 19 | 27 | 22 | 32 | 21 | 30 | **28** | **28** |
| Sawdust | % | | | 11 | 10 | 11 | 11 | 13 | 11 | 11 | | 10 | 13 | | | 11 | 12 | 12 | 12 | 10 | 10 | **11** | **11** |
| Shavings | % | | | 1 | | | 2 | 1 | | | | | 3 | | | | | | 2 | | 3 | **2** | **2** |
| Shrinkage loss | % | | | 3 | 2 | 5 | 4 | 2 | 1 | 5 | | 6 | 2 | | | 12 | 5 | 7 | 5 | 6 | 3 | **5** | **5** |
| Average sawnwood shipping weight | kg/m³ | | | | | | | | 415 | 477 | | 440 | 550 | | | 600 | 600 | | | | 549 | **549** | **519** |
| **NON-CONIFER** | Share (%) | 2 | 83 | 5 | 17 | 1 | 17 | 6 | 0 | 32 | 30 | 1 | 11 | 2 | | 27 | 13 | 37 | 1 | 26 | 2 | | |
| **Sawnwood all** | m³rw/m³p | 2.81 | 2.00 | 2.17 | 2.10 | 1.85 | 2.13 | 1.66 | 1.89 | 2.08 | 1.96 | 2.00 | 1.59 | 1.76 | | 1.52 | 1.55 | 1.89 | 1.90 | 1.62 | 2.50 | **1.90** | **1.95** |
| Sawnwood green rough | m³rw/m³p | | | | | | 2.20 | 1.66 | 1.89 | | 1.96 | 2.00 | 1.58 | | | 1.54 | 1.55 | 1.90 | | | | **1.89** | **1.81** |
| Sawnwood green planed | m³rw/m³p | | | | | | 2.85 | | | | | | | | | | | | | | | **2.85** | **2.85** |
| Sawnwood dry rough | m³rw/m³p | | | | | 1.85 | 2.50 | | | | | 2.20 | | | | 1.72 | | 2.10 | 1.90 | 1.91 | 2.50 | **2.01** | **2.09** |
| Sawnwood dry planed | m³rw/m³p | | | | | | 3.25 | | | | | | | | | | | | | | | **3.25** | **3.25** |
| Flitches and boules (rough green) | m³rw/m³p | | | | | | 1.60 | | | | | 1.40 | | | | 1.33 | | | | | | **1.40** | **1.44** |
| Flitches and boules (rough dry) | m³rw/m³p | | | | | | 1.75 | | | | | 1.70 | | | | 1.49 | | | | | | **1.70** | **1.65** |
| **Material balance** | | | | | | | | | | | | | | | | | | | | | | | |
| Sawnwood | % | 35 | | 46 | 47 | 54 | 47 | 60 | 53 | 48 | 51 | 50 | 63 | 57 | | 66 | | 53 | 53 | 61 | 40 | **53** | **52** |
| Chips/slabs | % | | | | 35 | | 36 | 13 | 35 | 36 | | 34 | 18 | | | 17 | | 27 | | 24 | | **31** | **28** |
| Sawdust | % | | | | 15 | | 10 | 13 | 11 | 11 | | 8 | 12 | | | 9 | | 14 | | 10 | | **11** | **11** |
| Shavings | % | | | | | | 4 | 7 | | | | | 5 | | | | | | | | | **5** | **5** |
| Shrinkage loss | % | | | | 3 | | 3 | 4 | 1 | 5 | | 8 | 2 | | | 8 | | 6 | | 5 | | **5** | **5** |
| Average sawnwood shipping weight | kg/m³ | | | | | | | | | 538 | | 550 | 700 | | | 790 | | | | | 699 | **699** | **655** |

**Notes:** m³rw = cubic metre roundwood, m³p = cubic metre product. / **Source**: FAO/ITTO/UNECE Forest product conversion factors questionnaire, 2018.

# CHAPTER 5

# 5. Veneer and plywood

Veneer is produced in one of two ways: by means of a lathe, in which the log is chucked and rotated against a stationary knife (rotary peeled veneer); or sliced, whereby the log is halved or quartered into flitches (sometimes called cants) with a saw and the flitch is then pressed against and moved across a knife (sliced veneer). Rotary peeling is generally used to produce thicker veneers for structural applications, and sliced veneer is used to produce thin veneers with decorative uses (but there are exceptions to these generalizations).

Plywood is a composite product manufactured by laminating sheets of veneer together into panels, generally with the layers of veneer oriented at 90 degrees to the previous layer (to improve multidirectional strength characteristics).

The determinants of veneer and plywood recovery ratios are similar to those for sawnwood: log size and characteristics, product specifications, and milling efficiency.

## 5.1 Volumetric and surface measurement

Two systems are generally used to measure veneer and plywood: surface measure, which strictly measures the surface area and does not account for volume (because thickness is not accounted for); and volumetric. Volumetric measures have two distinct variations: straight cubic volume (i.e. thickness × width × length); and surface measure on a thickness basis.

In most places, veneer and plywood is measured in m$^2$, m$^2$ 1 mm basis, and m$^3$. For example, a sheet of plywood that measures 1.22 m × 2.44 m × 12 mm has 35.72 m$^2$, 1 mm basis, which is easily converted to m$^3$ by dividing by 1 000 (= 0.03572 m$^3$).

In North America, veneer and plywood are often measured in square feet ⅜-inch basis. For example, a sheet of plywood measuring 4 feet × 8 feet × 0.472 inches has 40.37 ft$^2$ ⅜-inch basis (4 × 8 × [0.472 ÷ 0.375] = 40.37). Note that 0.375 is the decimal equivalent of the ⅜-inch basis thickness. To convert ft$^2$ ⅜ inch to cubic volume, divide by 1,130 to get m$^3$, or by 32 to get ft$^3$.

Typically, shrinkage has a significant effect on veneer and plywood because of the low moisture content (<6 percent) required.

## 5.2 Weight

As for sawnwood (see section 3.2), a theoretical approach can be used to calculate the weight of veneer. This will also work for plywood, but an allowance may need to be made for the weight of the glue line between the veneer plies. An approximate glue weight is 122 g per m$^2$ of glue-line surface area (United States Forest Service, 1956).

For example, an estimate is needed for the weight per volume (kg per m$^3$) for plywood made from Norway spruce measuring 1.22 m × 244 m × 0.013 m with five plies (i.e. four glue lines). The assumptions are that the wood will have a basic density of 380 kg per m$^3$, the moisture content of the wood will be 8 percent, and shrinkage will be 9 percent. Weight per m$^3$ of the wood component, therefore, is calculated as: 380 ÷ (1 – 0.09) × (1 + 0.08) = panel wood weight of 451 kg per m$^3$.

The volume of the panel is calculated as 1.22 × 2.44 × 0.013 = 0.0387 m$^3$. Thus, the wood weight for the panel is 0.387 × 451 = 17.45 kg.

There are four glue lines, each with a surface area of 2.98 $m^2$ for a total of 4 × 2.98 = 11.9 $m^2$ of glue line. The glue weight per panel is 122 × 11.9 = 1 453 g (1.45 kg); each panel weighs 1.45 kg glue + 17.45 kg wood = 18.9 kg; weight per $m^3$ is 18.9 ÷ 0.0387 = 488 kg/$m^3$.

## 5.3 Material balance

The material balance for the production of plywood and veneer is relatively complex because, to varying degrees, other solid wood products are also made during the production of veneer – that is, sawnwood in the production of sliced veneer, and peeler cores[9] in the production of rotary peeled veneer. Sawnwood is often a coproduct in the production of flitches for slicing; it is made from sections of the log that are unsuitable for veneer but adequate for sawnwood. Sawnwood is also manufactured in the slicing process from that portion of the flitch that is grasped and held by the arm that moves the flitch across the stationary slicing knife. This sawnwood is commonly referred to as backing board.

Peeler cores, which are a coproduct of the manufacture of rotary peeled veneer, are often sold as roundwood (e.g. fence posts), and they can be chipped. Some peeler cores, however, are large enough for the manufacture of sawnwood. Figure 5.1 shows the typical product recovery and material balance in a North American coniferous rotary plywood mill.

9 When a log is rotary peeled, it is often held and pivoted by lathe chucks. In this situation, the minimum diameter limit of the log at which veneer can no longer be peeled is usually determined by the diameter of the chucks. Thus, a lathe chuck with a diameter of 8 cm will produce peeler cores that are slightly larger in diameter (e.g. 8.5 cm) to prevent the veneer knife from coming into contact with the lathe chucks.

FIGURE 5.1 PRODUCT RECOVERY, BY SMALL-END LOG DIAMETER, IN A TYPICAL NORTH AMERICAN CONIFEROUS ROTARY PLYWOOD MILL

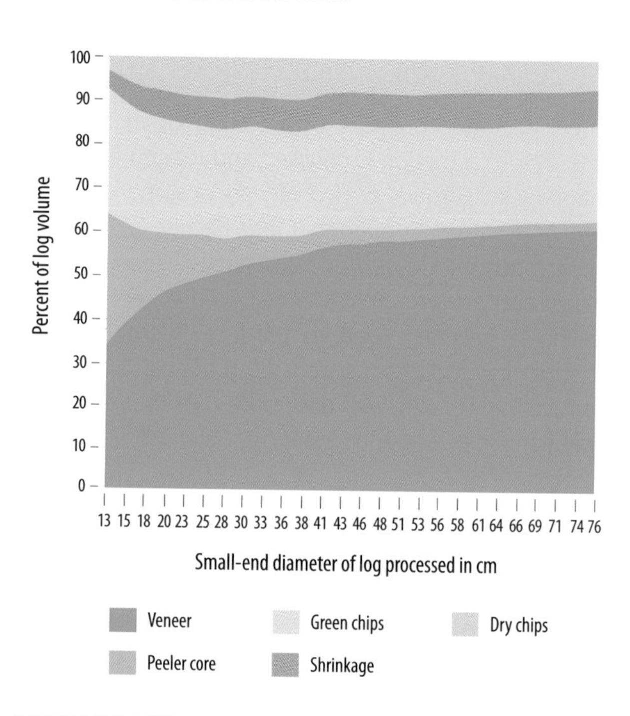


**Source:** Fonseca, 2005.

## 5.4 Summary of country data on veneer and plywood

### 5.4.1 Africa, Asia, Latin America, North America and Oceania

| | Unit in/ unit out | Africa | Asia | | | | | | Latin America | | | | | | | | North America | | | Oceania |
|---|---|---|---|---|---|---|---|---|---|---|---|---|---|---|---|---|---|---|---|---|
| | | South Africa | China | Indonesia | Japan | Malaysia | **Median** | **Average** | Brazil | Chile | Guyana | Mexico | Suriname | Uruguay | **Median** | **Average** | Canada | United States of America | **Median/ average** | New Zealand |
| **CONIFER** | Share (%) | 100 | 25 | 1 | 93 | | | | | 98 | | 70 | | 57 | | | 87 | 87 | | 100 |
| **(plywood and veneer)** | m³rw/m³p | 1.85 | 2.12 | 2.00 | 1.62 | | **2.00** | **1.91** | | 2.20 | | | | 2.38 | **2.29** | **2.29** | 1.92 | 1.84 | **1.88** | 1.79 |
| Rotary peeled veneer, green rough | m³rw/m³p | | 1.95 | 1.60 | | | **1.78** | **1.78** | | | | 1.60 | | | **1.60** | **1.60** | | 1.51 | **1.51** | |
| Rotary peeled veneer, dry rough | m³rw/m³p | | 2.25 | 1.90 | | | **2.08** | **2.08** | | 2.00 | | 1.80 | | | **1.90** | **1.90** | | 1.60 | **1.60** | |
| Rotary peeled plywood, dry rough | m³rw/m³p | | 1.90 | 1.90 | | | **1.90** | **1.90** | | 2.20 | | 2.00 | | | **2.10** | **2.10** | | 1.87 | **1.87** | |
| Rotary peeled plywood, dry sanded | m³rw/m³p | | 1.96 | 1.95 | | | **1.96** | **1.96** | | | | 2.10 | | | **2.10** | **2.10** | | 1.92 | **1.92** | |
| Sliced veneer, green rough | m³rw/m³p | | 1.55 | 1.60 | | | **1.58** | **1.58** | | | | 3.00 | | | **3.00** | **3.00** | | 1.51 | **1.51** | |
| Sliced veneer, dry rough | m³rw/m³p | | 1.65 | 1.75 | | | **1.70** | **1.70** | | 2.90 | | 3.20 | | | **3.05** | **3.05** | | 1.60 | **1.60** | |
| Sliced plywood, dry rough | m³rw/m³p | | 1.90 | 1.80 | | | **1.85** | **1.85** | | | | 4.20 | | | **4.20** | **4.20** | | 1.87 | **1.87** | |
| Sliced plywood, dry sanded | m³rw/m³p | | 1.95 | 1.90 | | | **1.93** | **1.93** | | | | 4.30 | | | **4.30** | **4.30** | | 1.92 | **1.92** | |
| **Material balance** | | | | | | | | | | | | 1.00 | | | | | | | | |
| Veneer | % | 53 | 60 | | | | **60** | **60** | | 50.0 | | 50 | | 42 | **50** | **47** | 52 | 53 | **53** | |
| Other products (chips, peeler cores, etc.) | % | 42 | 30 | | | | **30** | **30** | | | | 35 | | | **35** | **35** | 37 | 42 | **40** | |
| Sanding dust | % | 1 | 5 | | | | **5** | **5** | | | | 6 | | | **6** | **6** | 6 | 2 | **4** | |
| Shrinkage/losses | % | 4 | 5 | | | | **5** | **5** | | | | 9 | | | **9** | **9** | 5 | 3 | **4** | |
| Average panel shipping weight | kg/m³ | | | 620 | | | **620** | **620** | 420 | | | 550 | | 519 | **519** | **496** | | | | |
| Average panel thickness | mm | | | 15.0 | | | **15.0** | **15.0** | | | | 11.0 | | | **11.0** | **11.0** | | | | |
| **NON-CONIFER** | Share (%) | | 75 | 99 | 7 | 100 | | | | 2 | 100 | 30 | 100 | 43 | | | 13 | 13 | | |
| **(plywood and veneer)** | m³rw/m³p | | 2.00 | 2.10 | 1.62 | | **2.00** | **1.91** | | | | | 2.00 | 2.38 | **2.19** | **2.19** | 1.89 | 2.09 | **2.09** | |
| Rotary peeled veneer, green rough | m³rw/m³p | | 1.85 | 1.70 | | 2.00 | **1.85** | **1.85** | | | | 1.90 | | | **1.90** | **1.90** | | 2.00 | **2.00** | |
| Rotary peeled veneer, dry rough | m³rw/m³p | | 1.90 | 1.80 | | | **1.85** | **1.85** | | | | 2.10 | | | **2.10** | **2.10** | | 2.00 | **2.00** | |
| Rotary peeled plywood, dry rough | m³rw/m³p | | 1.90 | 1.95 | | 2.20 | **1.95** | **2.02** | | | | 2.20 | | | **2.20** | **2.20** | | 2.05 | **2.05** | |
| Rotary peeled plywood, dry sanded | m³rw/m³p | | 1.95 | 2.00 | | 2.25 | **2.00** | **2.07** | | | | 2.30 | | | **2.30** | **2.30** | | 2.14 | **2.14** | |
| Sliced veneer, green rough | m³rw/m³p | | 1.65 | 1.70 | | | **1.68** | **1.68** | | | | 3.00 | | | **3.00** | **3.00** | | 1.51 | **1.51** | |
| Sliced veneer, dry rough | m³rw/m³p | | 1.72 | 1.80 | | | **1.76** | **1.76** | | | | 3.20 | | | **3.20** | **3.20** | | 1.60 | **1.60** | |
| Sliced plywood, dry rough | m³rw/m³p | | 1.75 | 1.85 | | | **1.80** | **1.80** | | | | 4.20 | | | **4.20** | **4.20** | | 1.87 | **1.87** | |
| Sliced plywood, dry sanded | m³rw/m³p | | 1.80 | 1.95 | | | **1.88** | **1.88** | | | | 4.30 | | | **4.30** | **4.30** | | 1.92 | **1.92** | |
| **Material balance** | | | | | | | | | | | | | | | | | | | | |
| Veneer | % | | 65 | | | 50 | **58** | **58** | | | | 40 | | 42 | **41** | **41** | 53 | 48 | **51** | |
| Other products (chips, peeler cores, etc.) | % | | 30 | | | 38 | **34** | **34** | | | | 45 | | | **45** | **45** | 35 | 47 | **41** | |
| Sanding dust | % | | 3 | | | 4 | **4** | **4** | | | | 5 | | | **5** | **5** | 7 | 2 | **5** | |
| Shrinkage/losses | % | | 2 | | | 8 | **5** | **5** | | | | 10 | | | **10** | **10** | 5 | 3 | **4** | |
| Average panel shipping weight | kg/m³ | | | 650 | | 650 | **650** | **650** | 470 | | 764 | 700 | | 545 | **623** | **620** | | | | |
| Average panel thickness | mm | | | 21.0 | | 15.0 | **18.0** | **18.0** | | | | 5.0 | | | **5.0** | **5.0** | | | | |

**Notes:** m³rw = cubic metre roundwood, m³p = cubic metre product. / **Source**: FAO/ITTO/UNECE Forest product conversion factors questionnaire, 2018.

## 5.4.2 Europe

| | Unit in/ unit out | Europe | | | | | | | | | | | | | | | | |
|---|---|---|---|---|---|---|---|---|---|---|---|---|---|---|---|---|---|---|
| | | Croatia | Finland | France | Germany | Ireland | Lithuania | Poland | Portugal | Russian Federation | Slovakia | Slovenia | Spain | Sweden | Ukraine | UK | **Median** | **Average** |
| **CONIFER** | Share (%) | 0 | 70 | 33 | 100 | | | 31 | 63 | 5 | 36 | 5 | 40 | 100 | | | | |
| **(plywood and veneer)** | $m^3rw/m^3p$ | | 2.31 | 2.00 | | | | 1.95 | 2.47 | | | | | | 2.28 | | **2.28** | **2.20** |
| Rotary peeled veneer, green rough | $m^3rw/m^3p$ | | | 2.00 | 1.82 | | | | | | 2.00 | | 2.35 | | | | **2.00** | **2.04** |
| Rotary peeled veneer, dry rough | $m^3rw/m^3p$ | | | 1.90 | | | | 2.40 | | | 2.25 | | | 1.90 | | | **2.08** | **2.11** |
| Rotary peeled plywood, dry rough | $m^3rw/m^3p$ | | 2.27 | 2.00 | | | | | | | | | 1.85 | 2.50 | | | **2.14** | **2.16** |
| Rotary peeled plywood, dry sanded | $m^3rw/m^3p$ | | 2.31 | 2.00 | | | | | | | | | | | | | **2.16** | **2.16** |
| Sliced veneer, green rough | $m^3rw/m^3p$ | | | | | | | | | | 1.60 | | 5.20 | | | | **3.40** | **3.40** |
| Sliced veneer, dry rough | $m^3rw/m^3p$ | | | | | | | | | | 1.70 | | | | | | **1.70** | **1.70** |
| Sliced plywood, dry rough | $m^3rw/m^3p$ | | | | | | | | | | | | | | | | | |
| Sliced plywood, dry sanded | $m^3rw/m^3p$ | | | | | | | | | | | | | | | | | |
| **Material balance** | | | | | | | | | | | | | | | | | | |
| Veneer | % | | 43 | 50 | 55 | | | 51 | | | | | | | | | **51** | **50** |
| Other products (chips, peeler cores, etc.) | % | | 47 | 41 | | | | 41 | | | | | | | | | **41** | **43** |
| Sanding dust | % | | 3 | 4 | | | | 5 | | | | | | | | | **4** | **4** |
| Shrinkage/losses | % | | 7 | 5 | | | | 3 | | | | | | | | | **5** | **5** |
| Average panel shipping weight | $kg/m^3$ | | 535 | 550 | | 650 | | 650 | | | 500 | | | | | 649 | **600** | **589** |
| Average panel thickness | mm | | | | | 18.0 | | | | | 15.0 | | | | | | **16.5** | **16.5** |
| **NON-CONIFER** | Share (%) | 100 | 30 | 67 | | | | 69 | 37 | 95 | 64 | 95 | 60 | | | | | |
| **(plywood and veneer)** | $m^3rw/m^3p$ | 2.46 | 2.68 | 2.40 | | | | 1.95 | | 2.27 | | 2.50 | | | 1.77 | | **2.40** | **2.29** |
| Rotary peeled veneer, green rough | $m^3rw/m^3p$ | | | 2.40 | | | 2.17 | | | 1.62 | 1.95 | | 3.20 | | | | **2.17** | **2.27** |
| Rotary peeled veneer, dry rough | $m^3rw/m^3p$ | | | | | | | 2.30 | | 1.79 | 2.15 | | | | | | **2.15** | **2.08** |
| Rotary peeled plywood, dry rough | $m^3rw/m^3p$ | | 2.63 | | | | | | | 2.44 | 2.25 | | | | | | **2.44** | **2.44** |
| Rotary peeled plywood, dry sanded | $m^3rw/m^3p$ | | 2.68 | | | | | | | 2.54 | | | 2.40 | | | | **2.54** | **2.54** |
| Sliced veneer, green rough | $m^3rw/m^3p$ | | | | | | | | | | 1.60 | | 3.00 | | | | **2.30** | **2.30** |
| Sliced veneer, dry rough | $m^3rw/m^3p$ | | | | | | | | | | 1.70 | | | | | | **1.70** | **1.70** |
| Sliced plywood, dry rough | $m^3rw/m^3p$ | | | | | | | | | | | | | | | | | |
| Sliced plywood, dry sanded | $m^3rw/m^3p$ | | | | | | | | | | | | | | | | | |
| **Material balance** | | | | | | | | | | | | | | | | | | |
| Veneer | % | | 37 | 45 | | | 46 | 51 | | 44 | | 40 | | | | | **45** | **44** |
| Other products (chips, peeler cores, etc.) | % | | 54 | | | | 45 | 45 | | 38 | | 50 | | | | | **45** | **46** |
| Sanding dust | % | | 2 | | | | 3 | 2 | | 4 | | 2 | | | | | **2** | **3** |
| Shrinkage/losses | % | | 9 | | | | 6 | 2 | | 14 | | 8 | | | | | **8** | **8** |
| Average panel shipping weight | $kg/m^3$ | | 680 | | | 650 | | 650 | | 650 | 600 | | | | | | **650** | **646** |
| Average panel thickness | mm | | | | | 18.0 | | | | 12.0 | 25.0 | | | | | | **18.0** | **18.3** |

**Notes:** $m^3rw$ = cubic metre roundwood, $m^3p$ = cubic metre product. / **Source** : FAO/ITTO/UNECE Forest product conversion factors questionnaire, 2018.

CanWel Plus
MADE IN USA
Plum Creek
Ultra-Core TM
40 23/32
A
Plum Creek
60 15/32

# CHAPTER 6

# 6. Panels made of wood particles

Unlike panels made of veneer, panels made of wood particles (e.g. particle board, OSB and fibreboard) can have significantly different ratios of solid wood volume per product volume, depending on the wood from which they are produced and the target density of the panel. Wood particles can be pressed into panels that are more or less dense than the parent wood; thus, a cubic metre of product will seldom equal a cubic metre of solid wood equivalent. In addition, introduced non-wood components such as binders and fillers add to the bulk and weight of the product. In some countries, non-structural panels may include a significant percentage of bark volume (Thivolle-Cazat, 2008).

Many panel producers use the oven-dry weight of raw material, rather than the volume of solid wood input, to track the efficiency of raw-material-to-product conversion. The higher the wood density of the raw materials, the lower the volume of wood particles needed to make a given quantity of panels. For example, Norway spruce has a basic density of 380 kg/m$^3$: 2 m$^3$ of solid-wood equivalent raw material would therefore be required per m$^3$ of MDF panel pressed to a basic density of 760 kg/m$^3$. If, on the other hand, Siberian larch (with a basic density of 460 kg/m$^3$) is used as the raw material, only 1.65 m$^3$ would be needed (discounting the small effects of binders and fillers).

## 6.1 Volumetric and surface measurement

As for plywood and veneer, panels made of wood particles are typically measured by cubic volume, surface measure, and surface measure on a thickness basis. The surface measure is square area (e.g. m$^2$), irrespective of thickness; surface measure on a thickness basis (commonly 1 mm outside North America) refers to measures that establish thickness. In North America, various basis standards apply to panels made from wood particles and measured on a thickness basis, depending on the product, as follows:

- OSB — ⅜ inch (0.375 inch)
- Particle board and MDF — ¾ inch (0.75 inch)
- Hardboard — ⅛ inch (0.125 inch)
- Insulation board — ½ inch (0.5 inch).

## 6.2 Weight

The weight of panels made from wood particles varies according to the density of the parent wood, the density at which the wood fibre is pressed into the panel, moisture content (typically 6–8 percent – Briggs, 1994), and the weight of binders and fillers.

©Shutterstock

## 6.3 Summary of country data on panels made of wood particles

### 6.3.1 Asia, Latin America and North America

| | | *Asia* | | | | | *Latin America* | | | | | *North America* | | |
|---|---|---|---|---|---|---|---|---|---|---|---|---|---|---|
| | **Unit in/ unit out** | China | Indonesia | Malaysia | **Median** | **Average** | Brazil | Chile | Mexico | **Median** | **Average** | Canada | United States of America | **Median/ Average** |
| **Particle board (without OSB)** | m$^3$sw/m$^3$p | 1.60 | 1.60 | 1.40 | **1.60** | **1.53** | | 1.58 | 1.50 | **1.54** | **1.54** | 1.48 | 1.60 | **1.54** |
| Average thickness | mm | 18.0 | 18.0 | 18.0 | **18.0** | **18.0** | | 15.0 | 18.0 | **16.5** | **16.5** | 15.9 | | **15.9** |
| Product basic density | kg/m$^3$ | 660 | 700 | 700 | **700** | **687** | 640 | 640 | 720 | **640** | **667** | | 720 | **720** |
| **Material balance** | | | | | | | | | | | | | | |
| Binders and fillers | % | 10 | 9 | 8 | **9** | **9** | | 8 | 9 | **9** | **9** | | 9 | **9** |
| Bark | % | 3 | 1 | | **2** | **2** | | | 7 | **7** | **7** | | 1 | **1** |
| Moisture | % | 8 | 8 | 7 | **8** | **8** | | 8 | 7 | **8** | **8** | | 7 | **7** |
| Wood | % | 79 | 82 | 84 | **82** | **82** | | 84 | 77 | **81** | **81** | | 84 | **84** |
| Share of recycled fibre in panels | % | | | | | | | | | | | | | |
| **OSB and waferboard** | m$^3$sw/m$^3$p | 1.66 | 1.70 | | **1.68** | **1.68** | | 1.41 | | **1.41** | **1.41** | 1.61 | 1.65 | **1.63** |
| Average thickness | mm | 16.0 | 18.0 | | **17.0** | **17.0** | | 15.1 | | **15.1** | **15.1** | 11.1 | | **11.1** |
| Product basic density | kg/m$^3$ | 660 | 650 | | **655** | **655** | 640 | 705 | | **673** | **673** | | 407 | **407** |
| **Material balance** | | | | | | | | | | | | | | |
| Binders and fillers | % | 3 | 3 | 4 | **3** | **3** | | | | | | | 3 | **3** |
| Bark | % | 1 | 0 | | **1** | **1** | | | | | | | 1 | **1** |
| Moisture | % | 6 | 7 | 7 | **7** | **7** | | | | | | | 7 | **7** |
| Wood | % | 90 | 90 | 90 | **90** | **90** | | | | | | | 90 | **90** |
| **Fibreboard, hard (wet process)** | m$^3$sw/m$^3$p | 1.80 | 1.80 | | **1.80** | **1.80** | | 2.37 | 1.90 | **2.14** | **2.14** | 1.79 | 1.75 | **1.77** |
| Average thickness | mm | 3.00 | 3.00 | | **3.00** | **3.00** | | 3.00 | 3.00 | **3.00** | **3.00** | 3.20 | | **3.20** |
| Product basic density | kg/m$^3$ | 900 | 900 | | **900** | **900** | 940 | 960 | 850 | **940** | **917** | | 880 | **880** |
| **Material balance** | | | | | | | | | | | | | | |
| Binders and fillers | % | 7 | 9 | | **8** | **8** | | | | | | | | |
| Bark | % | 8 | 0 | | **4** | **4** | | | | | | | | |
| Moisture | % | 5 | 7 | | **6** | **6** | | 7 | | **7** | **7** | | | |
| Wood | % | 80 | 85 | | **83** | **83** | | | | | | | | |
| **Fibreboard, medium/high (MDF/HDF)** | m$^3$sw/m$^3$p | | 1.80 | | **1.80** | **1.80** | | 1.53 | | **1.53** | **1.53** | 1.45 | 1.60 | **1.53** |
| Average thickness | mm | | 16.0 | 12.0 | **14.0** | **14.0** | | 15.0 | 18.0 | **16.5** | **16.5** | 15.9 | | **15.9** |
| Product basic density | kg/m$^3$ | | 700 | 750 | **725** | **725** | 660 | 620 | 700 | **660** | **660** | | 704 | **704** |
| **Material balance** | | | | | | | | | | | | | | |
| Binders and fillers | % | | 9 | 9 | **9** | **9** | | 9 | | **9** | **9** | | | |
| Bark | % | | 0 | 0 | **0** | **0** | | 0 | 0 | **0** | **0** | | | |
| Moisture | % | | 7 | 6 | **7** | **7** | | 8 | 6 | **7** | **7** | | | |
| Wood | % | | 85 | 85 | **85** | **85** | | 83 | | **83** | **83** | | | |
| **Insulation board** | m$^3$sw/m$^3$p | | | | | | | | | | | | 0.71 | **0.71** |
| Average thickness | mm | | | | | | | | | | | | | |
| Product basic density | kg/m$^3$ | | | | | | | | | | | | 300 | **300** |
| **Material balance** | | | | | | | | | | | | | | |
| Binders and fillers | % | | | | | | | | | | | | | |
| Bark | % | | | | | | | | | | | | | |
| Moisture | % | | | | | | | | | | | | | |
| Wood | % | | | | | | | | | | | | | |
| Fibreboard, all | m$^3$sw/m$^3$p | | | | | | | 1.59 | | **1.59** | **1.59** | | | |

**Notes:** m$^3$sw = cubic metre solid wood; m$^3$p = cubic metre product; HDF = high-density fibreboard.

**Source:** FAO/ITTO/UNECE Forest product conversion factors questionnaire, 2018.

### 6.3.2 Europe

| | unit in/ unit out | Czech Republic | Finland | France | Germany | Ireland | Lithuania | Norway | Poland | Portugal | Slovakia | Slovenia | Spain | Sweden | Ukraine | United Kingdom | **Median** | **Average** |
|---|---|---|---|---|---|---|---|---|---|---|---|---|---|---|---|---|---|---|
| | | *Europe* | | | | | | | | | | | | | | | | |
| **Particle board (without OSB)** | m³sw/m³p | | 1.59 | 1.70 | 1.30 | 1.93 | 1.51 | 1.50 | 1.60 | 1.40 | 1.20 | | 1.40 | 1.50 | 1.80 | 1.62 | **1.51** | **1.54** |
| Average thickness | mm | 23.0 | | 19.0 | | 18.0 | | 18.0 | | 17.8 | 19.0 | | 16.0 | | 18.0 | | **18.0** | **18.6** |
| Product basic density | kg/m³ | 668 | 675 | 650 | | 625 | | 665 | 650 | 680 | 650 | | 650 | | 620 | 680 | **650** | **656** |
| **Material balance** | | | | | | | | | | | | | | | | | | |
| Binders and fillers | % | | 9 | 6 | | 8 | | 9 | 7 | 10 | 9 | | | | 9 | | **9** | **8** |
| Bark | % | | 5 | 5 | | 5 | | | 1 | 0 | | | | | 5 | | **5** | **4** |
| Moisture | % | | 7 | 7 | | 8 | | 7 | 7 | 7 | 8 | | | | 8 | | **7** | **7** |
| Wood | % | | 79 | 82 | | 79 | | 83 | 85 | 83 | 83 | | | | 88 | | **83** | **83** |
| Share of recycled fibre in panels | % | | | 25 | | 45 | | | 50 | 30 | | | | | | | **38** | **38** |
| **OSB and waferboard** | m³sw/m³p | 1.53 | | 1.90 | 1.30 | 1.93 | | | 1.85 | | | | | | | 1.58 | **1.72** | **1.68** |
| Average thickness | mm | 19.0 | | | | 18.0 | | | | | | | | | | 12.0 | **18.0** | **16.3** |
| Product basic density | kg/m³ | 565 | | 610 | | 630 | | | 650 | | | | | | | 620 | **620** | **615** |
| **Material balance** | | | | | | | | | | | | | | | | | | |
| Binders and fillers | % | | | 5 | | 4 | | | 3 | | | | | | | | **4** | **4** |
| Bark | % | | | 0 | | 0 | | | 10 | | | | | | | | **0** | **3** |
| Moisture | % | | | 7 | | 6 | | | 7 | | | | | | | | **7** | **7** |
| Wood | % | | | 88 | | 90 | | | 80 | | | | | | | | **88** | **86** |
| **Fibreboard, hard (wet process)** | m³sw/m³p | | | 2.23 | 2.40 | 1.93 | 2.03 | | 2.20 | 1.90 | | | 1.75 | | 2.30 | 2.37 | **2.20** | **2.12** |
| Average thickness | mm | | 3.00 | 3.00 | | 3.00 | | | | | | | 3.00 | | 3.20 | | **3.00** | **3.04** |
| Product basic density | kg/m³ | | 950 | 950 | 900 | 940 | | | 950 | | | | 830 | | 820 | 950 | **945** | **911** |
| **Material balance** | | | | | | | | | | | | | | | | | | |
| Binders and fillers | % | | | 0 | | 9 | | | 8 | | | | | | | | **8** | **6** |
| Bark | % | | | 3 | | 0 | | | | | | | | | | | **2** | **2** |
| Moisture | % | | | 6 | | 6 | | | 5 | | | | | | 8 | | **6** | **6** |
| Wood | % | | | 91 | | 85 | | | 87 | | | | | | | | **87** | **88** |
| **Fibreboard, medium/high (MDF/HDF)** | m³sw/m³p | | | 1.51 | 1.70 | 1.93 | | | 2.00 | | | 1.80 | 1.75 | | | 1.85 | **1.80** | **1.79** |
| Average thickness | mm | 24.0 | | 16.0 | | 15.0 | | | | 16.0 | | | 16.5 | | | | **16.0** | **17.5** |
| Product basic density | kg/m³ | 765 | | 750 | 650 | 740 | | | 700 | 740 | | 850 | 730 | | | 720 | **740** | **738** |
| **Material balance** | | | | | | | | | | | | | | | | | | |
| Binders and fillers | % | | | 9 | | 9 | | | 8 | 8 | | | | | | | **9** | **9** |
| Bark | % | | | 3 | | 0 | | | 10 | 0 | | | | | | | **2** | **3** |
| Moisture | % | | | 6 | | 6 | | | 3 | 7 | | | | | | | **6** | **6** |
| Wood | % | | | 82 | | 85 | | | 79 | 85 | | | | | | | **84** | **83** |
| **Insulation board** | m³sw/m³p | | | 0.60 | 1.10 | | | | 0.68 | | 1.10 | | | | | 0.63 | **0.68** | **0.82** |
| Average thickness | mm | | | 100.0 | | | | | | | 25.0 | | | | | | **62.5** | **62.5** |
| Product basic density | kg/m³ | 220 | 290 | 150 | 400 | | | | 300 | | 260 | | | | | 250 | **260** | **267** |
| **Material balance** | | | | | | | | | | | | | | | | | | |
| Binders and fillers | % | | | 5 | | | | | 5 | | 2 | | | | | | **5** | **4** |
| Bark | % | | | 0 | | | | | | | | | | | | | **0** | **0** |
| Moisture | % | | | 6 | | | | | 5 | | 6 | | | | | | **6** | **6** |
| Wood | % | | | 89 | | | | | 90 | | | | | | | | **90** | **90** |
| Fibreboard, all | m³sw/m³p | | 1.98 | | | | | | | | | 1.80 | | | | | **1.89** | **1.89** |

**Notes:** m³sw = cubic metre solid wood, m³p = cubic metre product; HDF = high-density fibreboard.

**Source:** FAO/ITTO/UNECE Forest product conversion factors questionnaire, 2018.

# CHAPTER 7

# 7. Wood pulp and paper

Pulp is the raw material for paper and paperboard (cardboard), as well as specialty pulps for the manufacture of textiles. The manufacturing process involves separating wood fibres by means of mechanical or chemical processes, or a combination of these.

Mechanical pulp is produced through a grinding action applied to wood fibre, typically wood particles. Because of the grinding, the wood fibres tend to be short and thus lack strength compared with chemically produced pulp. The yield is high, however, because little of the original wood fibre is lost. One oven-dry tonne of wood fibre input will yield approximately 0.95 tonnes of oven-dry mechanical pulp (i.e. 95 percent yield).

Chemical pulping involves the use of chemicals and heat to dissolve the lignin, leaving behind long wood fibres (which means relatively high strength) but at the cost of low yield compared with mechanical processes because much of the original wood fibre is dissolved and suspended in the chemical treatment. There are several chemical processes, the choice of which depends on the species comprising the source of the wood fibre and the desired characteristics of the paper. Bleached pulp produces white paper but the yield is lower compared with non-bleached pulp. Chemical pulping generally results in yields in the range of 40–50 percent (measured as oven-dry input to oven-dry output) (Briggs, 1994).

Pulp is also produced using mechanical grinding in conjunction with chemical processes. There are several variations of this process, and yield is generally slightly less than that of strictly mechanical processes. Table 7.1 shows typical yields for various pulp types, measuring pulp output in dry weight divided by the dry weight of the wood-fibre input.

## 7.1 Weight

Pulp and paper are usually measured by weight. Note that pulp and paper moisture content is assessed on a "wet basis" (mcw); generally, an unspecified tonne or an "air-dried" tonne is assumed to be 10 percent mcw, and 1 air-dried tonne of pulp is assumed, therefore, to comprise 900 kg of oven-dry fibre and 100 kg of contained water. Pulp and paper can also be measured on an oven-dry basis. For the purposes of the conversion factor questionnaire, conversions to tonnes were requested and respondents were informed that 10 percent mcw was assumed.

TABLE 7.1 TYPICAL PULP YIELDS FOR VARIOUS TYPES OF WOOD PULP

| *Pulp type* | *Pulp yield (% oven-dried output/input)* |
|---|---|
| **MECHANICAL** | |
| Mechanical (groundwood) | 93–95 |
| Thermomechanical | 80–90 |
| Chemithermomechanical | 80–90 |
| Chemimechanical | 80–90 |
| **SEMIMECHANICAL** | 70–85 |
| **CHEMICAL** | |
| Sulfate (kraft) | 45–55 |
| Sulfite | 40–50 |
| Magnefite | 45–55 |
| Soda | 40–50 |
| Soda-oxygen | 45–55 |
| Soda-anthraquinone | 45–55 |
| **DISSOLVING** | 35 |

**Notes:** The presentation of yield in this manner (oven-dry output/input) minimizes the variability that can occur when measuring input in volume, or input weight, with moisture. For example, generally less volume of wood from high-wood-density species is required to produce a given quantity of pulp than from low-wood-density species, and more wet chips than dry chips are needed to produce a given quantity of pulp.

**Source:** Briggs, 1994.

## 7.2 Summary of country data on wood pulp and paper

### 7.2.1 Africa, Asia, Latin America, North America and Oceania

| | | Africa | Asia | | | | | | Latin America | | | | | North America | | | Oceania |
|---|---|---|---|---|---|---|---|---|---|---|---|---|---|---|---|---|---|
| | Unit in/ unit out | South Africa | China | Indonesia | Japan | Malaysia | **Median** | **Average** | Chile | Mexico | Uruguay | **Median** | **Average** | Canada | United States of America | **Median/ Average** | New Zealand |
| **Wood pulp** | $m^3$sw/mt | | 3.80 | 4.00 | 3.30 | | **3.80** | **3.70** | | | 3.43 | **3.43** | **3.43** | | | | |
| Mechanical | $m^3$sw/mt | 2.55 | 2.50 | 2.50 | | | **2.50** | **2.50** | 2.72 | | | **2.72** | **2.72** | 2.39 | 2.51 | **2.45** | 2.00 |
| Basic density of wood input | kg/$m^3$ | 450 | 450 | 425 | | | **438** | **438** | 405 | | | **405** | **405** | | 466 | **466** | |
| Semi-chemical | $m^3$sw/mt | | 2.80 | 2.75 | | | **2.78** | **2.78** | | | | | | | | | 3.66 |
| Basic density of wood input | kg/$m^3$ | | 400 | 450 | | | **425** | **425** | | | | | | | | | |
| Chemical | $m^3$sw/mt | | 4.90 | 4.00 | | 4.19 | **4.19** | **4.36** | 4.26 | 4.00 | 3.43 | **4.00** | **3.90** | 5.22 | 3.35 | **4.29** | 3.66 |
| Basic density of wood input | kg/$m^3$ | | 520 | 450 | | 530 | **520** | **500** | 500 | 500 | 449 | **500** | **483** | | 503 | **503** | |
| Sulfate bleached | $m^3$sw/mt | | 5.10 | 4.45 | | 4.19 | **4.45** | **4.58** | 4.32 | 4.50 | 3.43 | **4.32** | **4.08** | | | | |
| Basic density of wood input | kg/$m^3$ | 550 | 450 | 450 | | 530 | **450** | **477** | 524 | 500 | | **512** | **512** | | | | |
| Sulfate unbleached | $m^3$sw/mt | 4.27 | 4.80 | 4.45 | | | **4.63** | **4.63** | 5.31 | | | **5.31** | **5.31** | | | | |
| Basic density of wood input | kg/$m^3$ | 450 | 410 | 420 | | | **415** | **415** | 405 | | | **405** | **405** | | | | |
| Sulfite bleached | $m^3$sw/mt | | 5.00 | 4.60 | | | **4.80** | **4.80** | | | | | | | | | |
| Basic density of wood input | kg/$m^3$ | | 450 | 450 | | | **450** | **450** | | | | | | | | | |
| Sulfite unbleached | $m^3$sw/mt | | 4.40 | 4.60 | | | **4.50** | **4.50** | | | | | | | | | |
| Basic density of wood input | kg/$m^3$ | | 390 | 440 | | | **415** | **415** | | | | | | | | | |
| Dissolving grades | $m^3$sw/mt | 6.54 | | 5.75 | | | **5.75** | **5.75** | | | | | | | | | |
| Basic density of wood input | kg/$m^3$ | 600 | | | | | | | | | | | | | | | |
| Recovered paper (input to output) | mt/mt | | 1.22 | 1.25 | | 1.30 | **1.25** | **1.26** | | | 1.20 | **1.20** | **1.20** | | | | |
| Share of recycled fibre in total pulp | % | 39 | 78 | 35 | | 89 | **78** | **67** | | | | | | | 38 | **38** | |
| Dry-solid organic black liquor / tonne chemical woodpulp | odmt/mt | | | | | | | | | | | | | | | | |
| **Paper and paperboard** | $m^3$sw/mt | | | 3.50 | | | **3.50** | **3.50** | | | | | | | | | |
| Newsprint | $m^3$sw/mt | | 3.20 | 2.80 | | | **3.00** | **3.00** | 2.80 | | | **2.80** | **2.80** | 2.50 | | **2.50** | 2.54 |
| Uncoated mechanical | $m^3$sw/mt | | 3.50 | 3.50 | | | **3.50** | **3.50** | | | | | | | | | |
| Coated paper | $m^3$sw/mt | | 4.50 | 3.90 | | | **4.20** | **4.20** | | | | | | | | | |
| Sanitary and household paper | $m^3$sw/mt | | 4.90 | 4.50 | | | **4.70** | **4.70** | | | | | | | | | |
| Packaging materials | $m^3$sw/mt | | 3.50 | 3.25 | | | **3.38** | **3.38** | | | | | | | | | |
| Case materials | $m^3$sw/mt | | 3.50 | 4.20 | | | **3.85** | **3.85** | | | | | | | | | |
| Folding boxboards | $m^3$sw/mt | | 3.80 | 4.00 | | | **3.90** | **3.90** | | | | | | | | | |
| Wrapping paper | $m^3$sw/mt | | 3.30 | 3.75 | | | **3.53** | **3.53** | | | | | | | | | |
| Other paper mainly for packaging | $m^3$sw/mt | | 3.30 | 3.75 | | | **3.53** | **3.53** | | | | | | | | | |
| Other paper and paperboard | $m^3$sw/mt | | 3.00 | 3.70 | | | **3.35** | **3.35** | | | | | | | | | 3.93 |

**Notes:** $m^3$sw = cubic metre solid wood; mt = tonne (in this case assumed air-dry – 10% moisture, wet basis).

**Source:** FAO/ITTO/UNECE Forest product conversion factors questionnaire, 2018.

### 7.2.2 Europe

| | Unit in/ unit out | Europe | | | | | | | | | | | | | | | | |
|---|---|---|---|---|---|---|---|---|---|---|---|---|---|---|---|---|---|---|
| | | Austria | Czechia | Finland | France | Germany | Netherlands | Norway | Poland | Portugal | Russian Federation | Slovakia | Slovenia | Spain | Sweden | United Kingdom | **Median** | **Average** |
| **Wood pulp** | $m^3sw/mt$ | 4.06 | | | | | | | | | 3.76 | 3.86 | 2.60 | | 3.70 | | **3.76** | **3.60** |
| Mechanical | $m^3sw/mt$ | | | 2.48 | 2.52 | 2.48 | 2.50 | 2.60 | 2.70 | | 2.88 | 2.50 | 2.60 | | 2.40 | 2.50 | **2.50** | **2.56** |
| Basic density of wood input | $kg/m^3$ | | | | 421 | | | 390 | | | 580 | | 405 | | | | **413** | **449** |
| Semi-chemical | $m^3sw/mt$ | | | 2.49 | 2.98 | 2.70 | | | | | 3.00 | 3.00 | | | 2.30 | 2.75 | **2.75** | **2.75** |
| Basic density of wood input | $kg/m^3$ | | | | 420 | | | | | | 600 | | | | | | **510** | **510** |
| Chemical | $m^3sw/mt$ | | | | 4.90 | 4.70 | | | 5.20 | 3.40 | 4.10 | | | | | | **4.70** | **4.46** |
| Basic density of wood input | $kg/m^3$ | | | | 440 | | | 400 | 500 | 500 | 635 | | | | | | **500** | **495** |
| Sulfate bleached | $m^3sw/mt$ | | 4.75 | 4.15 | 4.85 | | | 5.10 | | | 4.90 | 4.50 | | 2.98 | 4.80 | 4.50 | **4.75** | **4.50** |
| Basic density of wood input | $kg/m^3$ | | 445 | | 450 | | | 400 | 530 | | 640 | | | 553 | | | **490** | **503** |
| Sulfate unbleached | $m^3sw/mt$ | | 4.10 | 3.80 | 4.76 | | | 4.75 | | | 4.30 | | | | 4.50 | | **4.40** | **4.37** |
| Basic density of wood input | $kg/m^3$ | | 445 | | 420 | | | 400 | | | 560 | | | | | | **433** | **456** |
| Sulfite bleached | $m^3sw/mt$ | | 5.01 | | 5.08 | | | | 4.70 | | 4.30 | | | | 4.70 | 5.00 | **4.85** | **4.80** |
| Basic density of wood input | $kg/m^3$ | | 445 | | 410 | | | | 470 | | 575 | | | | | | **458** | **475** |
| Sulfite unbleached | $m^3sw/mt$ | | | | 4.88 | | | | | | 4.00 | | | | 4.40 | | **4.40** | **4.43** |
| Basic density of wood input | $kg/m^3$ | | | | 410 | | | | | | 575 | | | | | | **493** | **493** |
| Dissolving grades | $m^3sw/mt$ | | 6.27 | | | 5.10 | | | | | | | | | 6.20 | | **6.20** | **5.86** |
| Basic density of wood input | $kg/m^3$ | | 445 | | | | | | | | | | | | | | **445** | **445** |
| Recovered paper (input to output) | mt/mt | | | 1.30 | | 1.20 | | | 1.10 | 1.19 | 1.10 | 1.30 | 1.20 | 1.15 | | | **1.20** | **1.19** |
| Share of recycled fibre in total pulp | % | | | 5 | 57 | 84 | | | 46 | 45 | 34 | 24 | 62 | 69 | | | **46%** | **47%** |
| **Paper and paperboard** | $m^3sw/mt$ | | | | | | | | | 4.10 | 4.15 | 3.60 | | | | | **4.10** | **3.95** |
| Newsprint | $m^3sw/mt$ | | | 3.20 | | 3.20 | 2.80 | 2.70 | | | 3.15 | | | | | 2.80 | **2.98** | **2.98** |
| Uncoated mechanical | $m^3sw/mt$ | | | 3.50 | 3.18 | 3.50 | 3.50 | 2.60 | | | 4.00 | | | | | 3.50 | **3.50** | **3.40** |
| Coated paper | $m^3sw/mt$ | | | 4.40 | 3.40 | 4.40 | 3.50 | 2.50 | | | 4.00 | | | | | | **3.75** | **3.70** |
| Sanitary and household paper | $m^3sw/mt$ | | | 4.90 | 4.30 | 4.90 | 3.25 | | | | 4.20 | | | | | | **4.30** | **4.31** |
| Packaging materials | $m^3sw/mt$ | | | | | | 3.25 | | | | 4.30 | | | | | | **3.78** | **3.78** |
| Case materials | $m^3sw/mt$ | | | 4.20 | | 4.20 | 3.25 | | | | 4.00 | | | | | | **4.10** | **3.91** |
| Folding boxboards | $m^3sw/mt$ | | | 4.00 | | 4.00 | 3.25 | | | | 4.30 | | | | | | **4.00** | **3.89** |
| Wrapping paper | $m^3sw/mt$ | | | 4.10 | | 4.10 | 3.25 | | | | 4.40 | | | | | | **4.10** | **3.96** |
| Other paper mainly for packaging | $m^3sw/mt$ | | | 4.00 | | 4.00 | 3.25 | | | | 3.30 | | | | | | **3.65** | **3.64** |
| Other paper and paperboard | $m^3sw/mt$ | | | 3.70 | | 3.70 | 3.25 | | | | 3.30 | | | | | 2.50 | **3.30** | **3.29** |

**Notes:** $m^3sw$ = cubic metre roundwood; mt = tonne (in this case assumed air-dry – 10% moisture, wet basis).

**Source:** FAO/ITTO/UNECE Forest product conversion factors questionnaire, 2018.

# CHAPTER 8

# 8. Round and split wood products

For the purposes of this publication, this category of forest products includes barrel staves, utility poles, posts, pilings, house logs (manufactured roundwood for constructing log buildings) and shakes and shingles. Shingles are sawn rather than split, but they are included in this grouping because they are generally manufactured along with shakes, which are split. Similar to this, barrel staves are both split and sawn, depending on the characteristics of the wood species used.

These are relatively minor products in terms of volume, and information on them is limited. When no country data were reported (as was the case for shakes and shingles), information from previously published sources is provided.

This category of products is not as technology-driven as many other forest products; thus, this aspect does not appear to be a strong contributor to product recovery. Some of the significant determinants of product recovery are listed below, by product.

The yield of barrel staves (tapered wooden slats used in the cooperage industry) is determined, to a great degree, by the production process. Some species used for wine barrels (notably European oak species and Oregon white oak) need to be split to prevent the staves from being overly porous (caused by vessels that run parallel to the grain of the wood); on the other hand, white oak sourced from central and eastern North America does not have the same problem with porosity and can be sawn. The yield of sawn staves is significantly higher than that of split staves. Staves generally have exacting standards in terms of quality, reducing the yield according to the presence of knots and blemishes in the raw material used to produce them.

The yield for utility poles, posts, pilings and house logs is affected by peeling performed to remove bumps and protrusions and to round the profile. In addition, a portion of these products is peeled to remove taper, thereby creating a loss of 10–40 percent, depending on the diameter and length of the product, with smaller-diameter, longer products at the upper end of this range and larger-diameter, shorter products at the bottom.

Shakes and shingles are generally used for roofing and siding. Shakes are split on one or both faces and shingles are always sawn, thus losing a substantial volume to sawdust. Weatherization is an important factor, and quality standards tend to be exacting in terms of knots and other permeations that could allow water seepage. A substantial percentage of the original log volume is often unusable, therefore, and ends up as residue.

## 8.1 Volumetric and surface measurement

It is usually straightforward to convert from piece to volume for staves, poles, posts, pilings and house logs, which tend to be manufactured to exacting specifications.

This is not the case for shakes and shingles, however, which are typically measured on the basis of surface coverage. In North America, the unit is the "square", which is the quantity of shakes or shingles to cover 100 square feet (10.76 $m^2$). In Europe, the unit is square metres of coverage. Converting shakes and shingles from a surface (coverage) measure to a solid wood equivalent volume is not as straightforward because there are many combinations of thickness, taper,

TABLE 8.1 SHAKES AND SHINGLES PRODUCED FROM A CUBIC METRE OF WESTERN RED CEDAR LOGS (squares/m³ log volume)

| Log quality | Shingles | Shakes |
|---|---|---|
| Low-grade logs | 1.55 squares (14.4 m² coverage) | N/A |
| Medium-grade logs | 1.65 squares (15.3 m² coverage) | N/A |
| High-grade logs | 1.70 squares (15.8 m² coverage) | 1.9 squares (17.7 m² coverage) |

**Source:** Herring and Massie, 1989.

width, length and overlap, by product classification. Additionally, shake thickness is irregular because of the splitting process. Recovery of shakes and shingles per m³ log appear to range from about 14.4 m² to about 17.7 m² (1.55 squares to 1.9 squares) (table 8.1).

## 8.2 Weight

Like all other wood products, the weight-to-volume ratio is driven by the basic density, moisture content and shrinkage of the wood. The methodology outlined in section 4.2 can be used to calculate the theoretical weights of round and split wood products. It is important to consider, however, that many of these products are chemically treated to prevent fungal decay and attacks by wood-boring insects and molluscs, and wood treatment will add roughly 160 kg per m³ to poles and pilings (Hartman *et al.*, 1981). Thus, if the untreated weight of a wood component is 600 kg per m³, the same component, treated, would weigh about 760 kg per m³. Shakes and shingles are also often treated (to minimize decay and retard fire).

## 8.3 Material balance

**Barrel staves**. The manufacture of staves can produce residual products such as chips, sawdust, shavings and sawnwood. Some wood residue from stave production finds its way into the wine industry in the form of chips or sticks used to flavour wine in non-oak (e.g. stainless steel) barrels.

**Utility poles, posts, pilings and house logs**. Because these products generally cannot have any decay, crooked portions or overly large knots, the removal of such defects from the logs used as raw material reduces the yield of these products and increase the yield of secondary products such as chips and energy wood.

**Shakes and shingles**. Hartman *et al.* (1981) listed the material balance for shingles at roughly 40 percent shingles, 27 percent solid residue and 33 percent sawdust. Briggs (1994) cited an unpublished study showing a recovery of only 24 percent for shingles and 53 percent for shakes, with the remaining balance of both (76 percent and 47 percent, respectively) classified as residue. It is likely that the large differences in these material balances could be due to the use of high-grade versus lower-grade logs as a raw material, which may change over time due (for example) to competition from sawmills.

## 8.4 Summary of country data on round and split wood products

### 8.4.1 Africa, Asia, Latin America, North America and Oceania

| | | Africa | | | Asia | | | Latin America | | | North America | | |
|---|---|---|---|---|---|---|---|---|---|---|---|---|---|
| | **Unit in/ unit out** | Nigeria | South Africa | **Median/ Average** | China | Indonesia | **Median/ Average** | Guyana | Uruguay | **Median/ Average** | Canada | United States of America | **Median/ average** |
| Utility poles (round) | $m^3rw/m^3p$ | 1.74 | 1.35 | **1.55** | 1.2 | 1.30 | **1.25** | | 1.00 | **1.00** | | | |
| Posts (round) | $m^3rw/m^3p$ | 1.2 | | **1.2** | 1.6 | 1.50 | **1.55** | 1.7 | 1.1 | **1.40** | | | |
| Pilings (round) | $m^3rw/m^3p$ | | | | 1.65 | 1.50 | **1.58** | | | | | | |
| House logs (round) | $m^3rw/m^3p$ | | | | 1.65 | | **1.65** | | | | | | |
| Barrel staves | | | | | | | | | | | | | |
| Wood shingles | $m^3rw/m^2p$ | | | | | | | | | | 0.0654 | 0.0654 | **0.0654** |
| Wood shakes | $m^3rw/m^2p$ | | | | | | | | | | 0.0565 | 0.0565 | **0.0565** |

**Notes:** $m^3rw$ = cubic metre roundwood; $m^3p$ = cubic metre product; $m^2p$ = square metre product.

**Source:** FAO/ITTO/UNECE Forest product conversion factors questionnaire, 2018.

### 8.4.2 Europe

| | | Europe | | | | | | | | | | | |
|---|---|---|---|---|---|---|---|---|---|---|---|---|---|
| | **Unit in/ unit out** | Finland | France | Germany | Ireland | Lithuania | Norway | Poland | Slovakia | Slovenia | Spain | **Median** | **Average** |
| Utility poles (round) | $m^3rw/m^3p$ | 1.30 | 1.12 | 1.20 | 1.10 | | 1.15 | 1.18 | 1.67 | 1.15 | | **1.17** | **1.23** |
| Posts (round) | $m^3rw/m^3p$ | 1.50 | 1.11 | 1.20 | 1.15 | | 1.50 | 1.18 | 1.67 | | | **1.20** | **1.33** |
| Pilings (round) | $m^3rw/m^3p$ | 1.50 | 1.11 | 1.20 | | | | | 1.75 | | | **1.35** | **1.39** |
| House logs (round) | $m^3rw/m^3p$ | 1.70 | | | | 1.67 | 1.65 | | | | | **1.67** | **1.67** |
| Barrel staves | | | 4.25 | 5.0 | | | | | 2.25 | | 2.90 | **3.58** | **3.60** |
| Wood shingles | $m^3rw/m^2p$ | | | | | | | | | | | | |
| Wood shakes | $m^3rw/m^2p$ | | | | | | | | | 0.025 | | **0.025** | **0.025** |

**Notes:** $m^3rw$ = cubic metre roundwood; $m^3p$ = cubic metre product; $m^2p$ = square metre product.

**Source:** FAO/ITTO/UNECE Forest product conversion factors questionnaire, 2018.

CHAPTER

# 9

# 9. Energy wood products and properties

This category includes fuelwood, bark and chipped fuel, as well as manufactured products such as pellets, briquettes, charcoal and wood-based ethanol. Chapters 1 and 2 include information applicable to the analysis of conversion factors for fuelwood, wood residue and bark for energy.

The questionnaire did not include a category for conversion factors related to black liquor, a coproduct produced in the chemical pulping process (e.g. for paper and paperboard production). However, several sources indicate that 1.5 tonnes of black liquor solids (about 10 tonnes of weak black liquor) is produced for every 1 air-dried tonne of kraft pulp, which accounts for about two-thirds of the global virgin pulp production and 90 percent of the chemical pulp production (Tran and Vakkilainnen, 2008; Clay, 2008). The higher heating value of black liquor solids is about 14 GJ/tonne.

## 9.1 Volumetric measurement

Energy wood products can be measured in solid $m^3$ or loose $m^3$ (bulk $m^3$). In the case of fuelwood, loose $m^3$ is often differentiated from stacked $m^3$, with loose $m^3$ applicable to randomly placed wood and stacked to structures that fit neatly together. Section 1.1.3 outlines the measurement of stacked wood.

## 9.2 Weight

When moisture content is accounted for, the weight of energy products is likely to be the most reliable unit for understanding the energy-generating potential of a given quantity of wood. Moisture not only displaces the potential weight of combustible material but also consumes some of the energy because it is vaporized when the wood is combusted.

## 9.3 Energy values

The energy content of wood can be reflected in various measures, with assumptions made on the proportion of the energy content that realistically can be used (no process of using wood for energy is 100 percent efficient). The measurement unit specified in the conversion factors questionnaire was the joule, and energy values were requested in "higher heating values". Although such values are unachievable, they provide a consistent basis on which users can apply their own assumptions on conversion efficiency.

## 9.4 Summary of country data on energy wood products and properties

### 9.4.1 Asia, Latin America and North America

| | | *Asia* | | | | | | *Latin America* | | | | | *North America* |
|---|---|---|---|---|---|---|---|---|---|---|---|---|---|
| | **Unit in/unit out** | China | Indonesia | Japan | Malaysia | **Median** | **Average** | Brazil | Chile | Uruguay | **Median** | **Average** | United States of America |
| Fuelwood | | | | | | | | | | | | | |
| Conifer | Share (%) | | | | | | | | 12 | 1 | | | |
| (typical shipping weight) | $m^3$rw/tonne | | | | | | | | 2.47 | 1.81 | **2.14** | **2.14** | |
| Wood basic density (dry weight/green $m^3$) | kg/ solid $m^3$ | | | | | | | 480 | 405 | 550 | **480** | **478** | |
| Higher heating value | $m^3$ rw/GJ | | | | | | | | | | | | |
| Solid wood volume / stacked volume | % | | | | | | | | 65 | | **65** | **65** | |
| Non-conifer | Share (%) | | | | | | | | 88 | 99 | | | |
| (typical shipping weight) | $m^3$rw/tonne | | | | | | | | 1.61 | 1.54 | **1.58** | **1.58** | |
| Wood basic density (dry weight/green $m^3$) | kg/solid $m^3$ | | | | 800 | **800** | **800** | 520 | 620 | 650 | **620** | **597** | |
| Higher heating value | $m^3$ rw/GJ | | | | | | | | | | | | |
| Solid wood volume/stacked volume | % | | | | | | | | | | | | |
| Pellets | $m^3$sw/$m^3$p solid | | 2.50 | | | **2.50** | **2.50** | | | | | | |
| Solid wood $m^3$ per tonne pellets | $m^3$sw/mt p (5–10% mcw) | 2.20 | 2.30 | | 2.25 | **2.25** | **2.25** | | | | | | 2.09 |
| Solid wood input to bulk $m^3$ pellets | $m^3$sw/$m^3$p bulk | 1.50 | 1.45 | | | **1.48** | **1.48** | | 1.53 | | **1.53** | **1.53** | 1.44 |
| Product basic density (solid volume, oven dry) | kg/$m^3$ | 1 100 | 1 050 | | | **1 075** | **1 075** | | | | | | |
| Bulk density (loose volume, 5–10% mcw) | kg/$m^3$ | 660 | 710 | | 600 | **660** | **657** | | 650 | | **650** | **650** | 689 |
| Higher heating value (bulk volume) | $m^3$bulk/Gj | 0.08 | 0.08 | | | **0.08** | **0.08** | | | | | | 0.08 |
| Pressed logs and briquettes | $m^3$rw/odmt | | | | 2.25 | **2.25** | **2.25** | | | | | | |
| Solid wood input to tonne product | $m^3$sw/mt p (5–10% mcw) | | | | | | | | | | | | |
| Product basic density (solid volume, oven dry) | kg/$m^3$ | | | | | | | | | | | | |
| Bulk density (loose volume) | kg/$m^3$ | | | | | | | | | | | | |
| Higher heating value | $m^3$bulk/GJ | | 0.09 | | | **0.09** | **0.09** | | | | | | |
| Bark and chipped fuel | | | | | | | | | | | | | |
| Product basic density (solid volume, oven dry) | kg/$m^3$ | | 360 | | | **360** | **360** | | | | | | |
| Bulk density (loose volume at 50% mcw) | kg/$m^3$ | | | | 200 | **200** | **200** | | | | | | |
| Higher heating value | $m^3$ rw/GJ | | 0.09 | | | **0.09** | **0.09** | | | | | | |
| Charcoal (average 3–4% mcd) | $m^3$rw/tonne | | 6.00 | 7.40 | 5.00 | **6.00** | **6.13** | | 6.00 | | **6.00** | **6.00** | |
| Wood-based ethanol | $m^3$rw/kilolitre | | 7.30 | | | **7.30** | **7.30** | | | | | | 6.80 |

**Notes:** The ratios shown for wood pellets assumes only the wood fibre input into the manufacture of pellets and does not include any wood fibre that may have been burned by heat dryers to remove moisture from the wood fibre. Note that, as for all ratios involving roundwood, volume is measured underbark.

**Source:** FAO/ITTO/UNECE Forest product conversion factors questionnaire, 2018.

## 9.4.2 Europe

| | Unit in/unit out | Austria | Croatia | Czechia | Denmark | Finland | France | Germany | Ireland | Lithuania | Moldova | Netherlands | Norway | Poland | Slovakia | Slovenia | Spain | Sweden | Ukraine | United Kingdom | **Median** | **Average** |
|---|---|---|---|---|---|---|---|---|---|---|---|---|---|---|---|---|---|---|---|---|---|---|
| | | Europe | | | | | | | | | | | | | | | | | | | | |
| Fuelwood | | | | | | | | | | | | | | | | | | | | | | |
| Conifer | Share (%) | | 4% | | 67% | | 10% | 40% | | | | 15% | 4% | 51% | | 15% | | | | | | |
| (typical shipping weight) | $m^3$rw/tonne | | | | | | 2.05 | | | | | 1.28 | 1.25 | | | 2.00 | | | | | **1.64** | **1.65** |
| Wood basic density (dry weight/green $m^3$) | kg/ solid $m^3$ | | 382 | 482 | 390 | | 407 | 389 | | | | 400 | 380 | | | 400 | | | | 462 | **400** | **410** |
| Higher heating value | $m^3$ rw/GJ | | | 0.105 | | | 0.13 | | | | | | 0.12 | | | 0.11 | | | | 0.13 | **0.12** | **0.12** |
| Solid wood volume / stacked volume | (%) | | 69% | | 65% | | 71% | | | | | | 65% | | | 65% | | | | | **65%** | **67%** |
| Non-conifer | Share (%) | | 96% | | 33% | | 90% | 60% | | | | 85% | 96% | 49% | | 85% | | | | | | |
| (typical shipping weight) | $m^3$rw/tonne | | 1.00 | | | | 1.50 | | | | | 1.06 | 1.11 | | | 1.40 | | | | | **1.11** | **1.21** |
| Wood basic density (dry weight/green $m^3$) | kg/ solid $m^3$ | | 564 | | 560 | | 550 | 563 | | | | 584 | 500 | | | 580 | | | | | **563** | **557** |
| Higher heating value | $m^3$ rw/GJ | | | | | | 0.1 | | | | | | 0.1 | | | 0.08 | | | | | **0.10** | **0.09** |
| Solid wood volume / stacked volume | % | | 69% | | | | | | | | | | 65% | | | 65% | | | | | **65%** | **66%** |
| Pellets | $m^3$sw/$m^3$p solid | | | 2.00 | | 2.86 | 2.86 | | 2.20 | 2.30 | | | | | 2.23 | 2.20 | | | | | **2.23** | **2.38** |
| Solid wood $m^3$ per tonne pellets | $m^3$sw/mt p (5–10% mcw) | 2.21 | 1.82 | | | 2.32 | 2.39 | 2.29 | | 2.16 | | | 2.27 | 2.00 | | 2.10 | | | 2.36 | | **2.24** | **2.19** |
| Solid wood input to bulk $m^3$ pellets | $m^3$sw/$m^3$p bulk | 1.44 | | | | 1.51 | 1.79 | 1.50 | | | | | | | | 1.34 | | | 1.53 | | **1.51** | **1.52** |
| Product basic density (solid volume, oven dry) | kg/$m^3$ | | | 1 100 | | 1 080 | 1 200 | 1 120 | 920 | 1 067 | | | | 1 120 | 1 070 | 1 100 | 1 010 | | 1 100 | | **1 100** | **1 081** |
| Bulk density (loose volume, 5–10% mcw) | kg/$m^3$ | 652 | | | | 650 | 650 | 650 | | | | | 650 | 600 | 670 | 640 | | | 650 | 677 | **650** | **649** |
| Higher heating value (bulk volume) | $m^3$bulk/GJ | 0.083 | | 0.060 | | 0.090 | 0.070 | 0.090 | 0.130 | | | | | | 0.080 | 0.080 | 0.070 | | 0.082 | 0.080 | **0.080** | **0.083** |
| Pressed logs and briquettes | $m^3$rw/odmt | 2.38 | | | | | 2.38 | | 2.20 | 2.32 | | | | | | 2.20 | | | | | **2.32** | **2.30** |
| Solid wood input to tonne product | $m^3$sw/mt p (5–10% mcw) | | | | | | 2.00 | 1.96 | | | | | 1.54 | | | | | | | | **1.96** | **1.83** |
| Product basic density (solid volume, oven dry) | kg/$m^3$ | | | 1 000 | | 1 080 | 1 000 | 1 200 | 950 | 1 075 | | | | | 1 120 | 1 100 | 1 100 | | | | **1 080** | **1 069** |
| Bulk density (loose volume) | kg/$m^3$ | 761 | | | | | 900 | | | | | | | 1 200 | | | | | | | **900** | **954** |
| Higher heating value | $m^3$bulk/GJ | 0.07 | | 0.06 | | 0.09 | 0.13 | | 0.13 | | | | | | 0.05 | | | | | | **0.080** | **0.088** |
| Bark and chipped fuel | | | | | | | | | | | | | | | | | | | | | | |
| Product basic density (solid volume, oven dry) | kg/$m^3$ | 393 | | 465 | | 400 | 400 | | | | | 443 | 380 | 300 | | 450 | | 350 | | 152 | **397** | **373** |
| Bulk density (loose volume at 50% mcw) | kg/$m^3$ | 236 | | | | | 300 | | | | | | 315 | | | 280 | | | | 192 | **280** | **265** |
| Higher heating value | $m^3$ rw/GJ | 0.12 | | | | | 0.13 | | | | | | 0.13 | | 0.08 | 0.10 | | | | | **0.11** | **0.12** |
| Charcoal (average 3-4% mcd) | $m^3$rw/tonne | | 5.70 | | | 6.10 | 7.00 | 5.00 | | | 5.50 | | 6.00 | 6.40 | 5.70 | | | | | 6.00 | **6.00** | **5.93** |
| Wood-based ethanol | $m^3$rw/kilolitre | | | | | | | 8.62 | | | | | | | | | | | | | **8.62** | **8.62** |

**Notes:** The ratios shown for wood pellets assumes only the wood fibre input into the manufacture of pellets and does not include any wood fibre that may have been burned by heat dryers to remove moisture from the wood fibre. Note that, as for all ratios involving roundwood, volume is measured underbark.

**Source:** FAO/ITTO/UNECE Forest product conversion factors questionnaire, 2018.

# References


Brandt, J.P., Morgan, T.A., Keegan, C.E., III, Songster, J.M., Spoelma, T.P. & DeBlander, L.T. 2012. *Idaho's forest products industry and timber harvest, 2006*. Resource Bulletin RMRS-RB-12. Fort Collins, USA, United States Department of Agriculture (available at www.bber.umt.edu/FIR/..%5Cpubs%5CForest%5Cfidacs%5CID2006.pdf).

Briggs, D. 1994. *Forest products measurements and conversion factors with special emphasis on the U.S. Pacific Northwest*. Seattle, USA, College of Forest Resources, University of Washington.

CITES. 2008. Convention on International Trade in Endangered Species of Wild Fauna and Flora, *Timber issues - Bigleaf mahogany - Volumetric conversion of standing trees to exportable mahogany sawn wood*. Geneva, Switzerland (available at www.cites.org/sites/default/files/eng/com/pc/17/E-PC17-16-01-03.pdf).

Clay, D. 2008. *Evaporation principles and black liquor properties*. TAPPI Press (available at www.tappi.org/content/events/08kros/manuscripts/3-1.pdf).

FAO. 1947. *Report of Subcommittee on Units of Measurements*. Conference on Forest Statistics, 11–14 February 1947. Rome (available at www.unece.org/fileadmin/DAM/timber/other/FAO-conversion-factors-1947.pdf).

Fonseca, M.A. 2005. *The measurement of roundwood: methodologies and conversion ratios*. Wallingford, UK, CABI Publishing.

Freese, F. 1973. *General Technical Report FPL-01*. Madison, USA, United States Department of Agriculture (available at www.fpl.fs.fed.us/documnts/fplgtr/fplgtr01.pdf).

Hartman, D.A., Atkinson, W.A., Bryant, B.S. & Woodfin, R.O. Jr. 1981. *Conversion factors for the Pacific Northwest Forest industry*. Seattle, USA, Institute of Forest Resources, College of Forest Resources, University of Washington.

Herring, R. & Massie, M. 1989. *Analysis of the British Columbia shake and shingle industry*. FRDA report 070. Victoria, Canada.

Mantau, U. 2008. *Presentation at the 2008 UNECE/FAO Workshop on National Wood Resources Balances* (available at www.unece.org/fileadmin/DAM/timber/workshops/2008/wood-balance/presentations/01_Mantau.pdf).

Matthews, G. 1993. *The carbon content of trees*. Forestry Commission Technical Paper 4. Edinburgh, UK. Forestry Commission (available at www.forestresearch.gov.uk/documents/6904/FCTP004.pdf).

McIver, C.P, Meek, J.P., Scudder, M.G, Sorenson, C.B, Morgan, T.A. & Christensen, G.A. 2015. *California's Forest Products Industry and Timber Harvest, 2012*. Portland, USA, United States Forest Service (available at www.fs.fed.us/pnw/pubs/pnw_gtr908.pdf).

National Hardwood Lumber Association. 1994. *Rules for the measurement and inspection of hardwood and cypress*. Memphis, USA, National Hardwood Lumber Association.

Nova Scotia Department of Natural Resources. 2001. *Conversion factors*. Halifax, Canada.

Ontario Ministry of Natural Resources. 2000. *Scaling Manual (Second Edition)*. Queen's Printer for Canada. Ontario, Canada

SDC. 2014. *SDC's instruction for timber measurement: measurement of roundwood stacks* [online]. Biometria. [Cited July 2019]. https://www.sdc.se/admin/Filer/Nya%20m%C3%A4tningsinstruktioner%202014/SDCs%20instructions%20-%20Measurement%20of%20roundwood%20stacks.pdf

South East Asia Lumber Producers' Association. 1981. *Log grading rules*. Approved by the SEALPA Council's 13th Meeting in Port Moresby April 14, 1981 (available at www.fiapng.com/Sealpa%20Log%20Grading%20Rules.pdf).

Spelter, H. 2002. *Conversion of board foot scaled logs to cubic meters in Washington State, 1970–1998*. General Technical Report FPL-GTR-131. Madison, USA, United States Forest Service.

Thivolle-Cazat, A. 2008. *Conversion factors: a necessity for an accurate estimation of wood consumption by industries. UNECE/FAO. Geneva, Switzerland* (available at www.unece.org/fileadmin/DAM/timber/workshops/2008/wood-balance/presentations/04_Thivolle-Cazat.pdf).

Tran, H. & Vakkilainnen, E.K. 2008. *The kraft chemical recovery process* (available at www.tappi.org/content/events/08kros/manuscripts/1-1.pdf).

UNECE/FAO. 2010a. *Forest products conversion factors for the UNECE region* (available at www.unece.org/fileadmin/DAM/timber/publications/DP-49.pdf).

UNECE/FAO. 2010b. *North American log scaling methods* [PowerPoint presentation]. Nordic Wood Measurement Meeting. Drammen, Norway, 26–27 October 2010.

United States Forest Service. 1956. *Estimating the Weight of Plywood* (technical note number 260). Washington, DC, United States Department of Agriculture..

United States Forest Service. 1999. *Wood handbook — wood as an engineering material.* General Technical Report FPL–GTR–113. Madison, USA.

Van Vuuren N., Banks, C. & Stöhr, H. 1978. *Shrinkage and density of timbers used in the Republic of South Africa.* Bulletin 57. Pretoria, Department of Forestry.

VMF Nord. 1999. *Estimation of the solid volume percentage.* Circular A 13. Swedish Timber Measurement Council.

Western Wood Products Association. 1998. *Western lumber grading rules.* Portland, USA, Western Wood Products Association.

## Roundwood measurement standards references

Alberta Ministry of Agriculture and Forestry. 2006. *Alberta scaling manual.* Edmonton, Canada (available at https://open.alberta.ca/publications/alberta-scaling-manual#summary).

Association pour la rationalisation et la mécanisation de l'explotation forestière, Centre technique du bois et de l'ameublement, Mutualité sociale agricole. 1994. *Manuel d'exploitation forestière, tome 2.* Fontainebleau, France, Centre technique du bois et de l'ameublement.

British Columbia Ministry of Ministry of Forests, Lands and Natural Resource Operations. 2018. *Scaling manual.* Victoria, Canada, Crown Publications (available at www2.gov.bc.ca/gov/content/industry/forestry/competitive-forest-industry/timber-pricing/timber-scaling/timber-scaling-manual).

Finland Ministry of Agriculture and Forestry. 1997. *Roundwood volume measurement standards* [Reg 918/66/97].

Forest Service of the Department of the Marine and Natural Resources. 1999. *Timber measurement manual: standard procedures for the measurement of round timber for sales purposes in Ireland* (available at www.coford.ie/media/coford/content/publications/projectreports/TimberMeasurementManual.pdf).

Government of Japan. 1967. *Japanese Agricultural Standard* (Public Notice No. 1841) (available at interpine.nz/wp-content/uploads/2011/08/JAS-Log-Scaling-Pubic-Notice-1841-MAF-Dec-8-1967.pdf).

Government of the Russian Federation. Undated. *GOST-Standard 2708-75 round timber* (volumes tables).

Hamilton, G.J. 1975. *Forest mensuration handbook.* London, Forestry Commission.

Northwest Log Rules Advisory Group. 2003. *Official log scaling and grading rules.* Eugene, USA.

Nova Scotia Department of Natural Resources. 2007. *Nova Scotia Scaling Manual 2nd Edition* (available at www.novascotia.ca/natr/forestry/scaling/pdf/ScalingManual.pdf).

Ontario Ministry of Natural Resources. 2017. *Scaling Manual Fourth Edition.* Queen's Printer for Canada. Ontario, Canada (available at https://files.ontario.ca/mnrf-forestry-scaling-manual-english-only-20190506.pdf).

Papua New Guinea Forest Authority. 1996. *Procedures for the identification, scaling and reporting (including royalty self-assessment) on logs harvested from natural forest logging operations.* Hohola, Papua New Guinea, Papua New Guinea Forest Authority.

Sächsisches Staatsministerium. 1997. *Messung und Sortierung von Roholz.* Dresden, Germany.

SDC. 2014a. *SDC's instruction for timber measurement: measurement of roundwood stacks* [online]. Biometria. [Cited July 2019]. www.biometria.se/instruktioner-for-matning-och-kontroll-av-matning

SDC. 2014b. *SDC's Instruction for timber measurement: measurement of log volume under bark.* Sundsvall, Sweden (available at: www.sdc.se/admin/Filer/Nya%20m%C3%A4tningsinstruktioner%202014/SDCs%20instructions%20-%20Measurement%20of%20log%20volume.pdf).

South East Asia Lumber Producers' Association. 1981. *Log grading rules.* Approved by the SEALPA Council's 13th Meeting in Port Moresby April 14, 1981 (available at http://www.fiapng.com/Sealpa%20Log%20Grading%20Rules.pdf).

Swedish Timber Measurement Council. 1999. *Regulations for measuring of roundwood.* Circular VMR 1-99. Sweden.

Tømmermålingforeningenes Fellesorgan. 1998. *Grading and scaling regulations for forestry products.* Norway.

Forestry Commission. 2006. *Forest mensuration: a handbook for practitioners.* Forestry Commission Publications. Cheshire, UK.

United States Forest Service. 1991. *National forest cubic scaling handbook.* FSH 2409.11a. Washington, DC (available at www.fs.fed.us/cgi-bin/Directives/get_directives/fsh?2409.11a).

02596
9524
2877

# Annex

ANNEX TABLE 1 LIST OF EQUIVALENTS

| To convert from | To | Multiply by |
|---|---|---|
| **Length** | | |
| millimetre | centimetre | 0.1 |
| millimetre | inch | 0.0394 |
| millimetre | foot | 0.00328 |
| millimetre | metre | 0.001 |
| centimetre | millimetre | 10 |
| centimetre | inch | 0.394 |
| centimetre | foot | 0.0328 |
| centimetre | metre | 0.01 |
| inch | millimetre | 25.4 |
| inch | centimetre | 2.54 |
| inch | foot | 0.0833 |
| inch | metre | 0.0254 |
| foot | millimetre | 304.8 |
| foot | centimetre | 30.48 |
| foot | inch | 12 |
| foot | metre | 0.3048 |
| metre | millimetre | 1,000 |
| metre | centimetre | 100 |
| metre | inch | 39.37 |
| metre | foot | 3.281 |
| **Area** | | |
| square centimetre | square inch | 0.155 |
| square centimetre | square foot | 0.001076 |
| square centimetre | square metre | 0.0001 |
| square inch | square centimetre | 6.452 |
| square inch | square foot | 0.0069444 |
| square inch | square metre | 0.0006452 |
| square foot | square centimetre | 929 |
| square foot | square inch | 144 |
| square foot | square metre | 0.0929 |
| square metre | square centimetre | 10 000 |
| square metre | square inch | 1 550 |
| square metre | square foot | 10 764 |
| **Density** | | |
| specific gravity | basic density | 1 000 |
| basic density | specific gravity | 0.001 |
| **Stacked measure** | | |
| cubic foot | stere | 0.02832 |
| cubic foot | cord | 0.0078125 |
| cubic metre | stere | 1 |
| cubic metre | cord | 0.2759 |
| cord | cubic foot | 128 |
| cord | cubic metre | 3.6245 |
| cord | stere | 3.6245 |
| stere | cubic foot | 35.315 |
| stere | cubic metre | 1 |
| stere | cord | 0.2759 |

| To convert from | To | Multiply by |
|---|---|---|
| **Weight** | | |
| gram | pound | 0.002205 |
| gram | kilogram | 0.001 |
| pounds | gram | 454 |
| pounds | kilogram | 0.454 |
| pounds | ton | 0.0005 |
| pounds | tonne | 0.000454 |
| kilogram | gram | 1000 |
| kilogram | pounds | 2.205 |
| kilogram | ton | 0.0011023 |
| kilogram | tonne | 0.001 |
| ton | kilogram | 907 |
| ton | pound | 2000 |
| ton | tonne | 1.1023 |
| tonne | pounds | 2 205 |
| tonne | kilogram | 1 000 |
| tonne | ton | 0.907 |
| **Volume** | | |
| square foot (⅜ inch basis) | square metre (1 mm basis) | 0.885 |
| square foot (⅜ inch basis) | cubic foot | 0.03125 |
| square foot (⅜ inch basis) | cubic metre | 0.000885 |
| square foot (⅜ inch basis) | cunit | 0.0003125 |
| square metre (1 mm basis) | square foot (⅜ inch basis) | 1.13008 |
| square metre (1 mm basis) | cubic foot | 0.35315 |
| square metre (1 mm basis) | cubic metre | 0.001 |
| square metre (1 mm basis) | cunit | 0.00035315 |
| cubic foot | square foot ⅜ | 32 |
| cubic foot | square metre (1 mm basis) | 28.32 |
| cubic foot | cubic metre | 0.02832 |
| cubic foot | cunit | 0.01 |
| cubic metre | square foot (⅜ inch basis) | 1 130.08 |
| cubic metre | square metre (1 mm basis) | 1 000 |
| cubic metre | cubic foot | 35.315 |
| cubic metre | cunit | 2.832 |
| cunit | square foot (⅜ inch basis) | 3 200 |
| cunit | square metre (1 mm basis) | 2 832 |
| cunit | cubic foot | 100 |
| cunit | cubic metre | 2.832 |
| **Heat** | | |
| British thermal unit | joule | 1 055 |
| joule | British thermal unit | 0.0009479 |
| **Weight to volume ratios** | | |
| pounds per cubic foot | kg per cubic metre | 16.019 |
| kg per cubic metre | pounds per cubic foot | 0.06243 |

**Source:** Fonseca, 2005.

**ANNEX FIGURE 1** EXAMPLE OF WOOD BALANCE USING CONVERSION FACTORS

1 593 Roundwood needed (1 000 $m^3$)

| Sector / item | Input (1 000 $m^3$) | Material balance (Conversion Factors) | Primary product (Forest Products, 1 000 $m^3$ swe) | Chips | Sawdust, shavings, sanding dust | Losses | Bark* | Total excluding bark | swe $m^3$/ product unit | Quantity (Products in unit of output) | Product unit (1 000) |
|---|---|---|---|---|---|---|---|---|---|---|---|
| **Sawmilling Sector** – Roundwood in | 1,000 | | | | | | | | | | |
| Sawnwood | | 50% (Σ) | 530 | | | | | | 1.06 | 500 | $m^3$ |
| Shrinkage | | 3% (Σ) | | | | | | | | | |
| Chips | | 33% | | 330 | | | | | 2.4 | 138 | odmt |
| Sanding and sawdust | | 10% | | | 100 | | | | 2.4 | 42 | odmt |
| Loss | | 4% | | | | 40 | | 1000 | | | |
| Recoverable bark | | 7% | | | | | 73 | | 2.6 | 28 | odmt |
| **Plywood/Veneer Mill** – Roundwood in | 200 | | | | | | | | | | |
| Plywood and veneer | | 53% (Σ) | 113 | | | | | | 1.08 | 105 | $m^3$ |
| Shrinkage | | 4% (Σ) | | | | | | | | | |
| Chips | | 37% | | 75 | | | | | 2.4 | 31.1 | odmt |
| Sanding and sawdust | | 3% | | | 6 | | | | 2.4 | 2.5 | odmt |
| Loss | | 3% | | | | 6 | | 200 | | | |
| Recoverable bark | | 7% | | | | | 15 | | 2.6 | 5.6 | odmt |
| **OSB Mill** – Roundwood in | 40 | | | | | | | | | | |
| Panels | | 60% (Σ) | 39.4 | | | | | | 1.64 | 24 | $m^3$ |
| Shrinkage (densification) | | 69% (Σ) | | | | | | | | | |
| Sanding and sawdust | | 1% | | | 0.5 | | | | 2.4 | 0.2 | odmt |
| Loss | | 0.3% | | | | 0.1 | | 40 | | | |
| Recoverable bark | | 7% | | | | | 3 | | 2.6 | 1.1 | odmt |
| **Pulp and paper Mill** – Roundwood in | 353 | | | | | | | | | | |
| Chips in | 405 | | | | | | | | | | |
| Pulp and paper | | 95% | 720 | | | | | | 2.4 | 300 | $m^3$ |
| Loss | | 5% | | | | 38 | | 758 | | | |
| Recoverable bark | | 7% | | | | | 26 | | 2.6 | 9.9 | odmt |
| **Particle board Mill** – Sawdust in | 106 | | | | | | | | | | |
| Particle board | | 67% (Σ) | 106.5 | | | | | 106.5 | 1.5 | 71 | $m^3$ |
| Shrinkage (densification) | | 33% (Σ) | | | | | | | | | |
| **Biomass power plant*** – Bark in | 116 | | | | | | | | | | |
| Power out | | 100% | 116 | | | | | 116 | 0.1824 | 636 | gj |

Energy out

1 509 $m^3$ Wood products

84 $m^3$ Losses

**Notes:** odmt = oven-dry tonne; swe = solid wood equivalent. Conversion factors are used in outlook studies to predict the amount of raw material needed from the forest to supply a given quantity of manufactured forest products and vice-versa. A wood balance can check the accuracy of conversion factors and assist in predicting the flow and availability of wood residues. This figure is an example of a simple wood balance using conversion factors and could represent a national or subregional balance. In the example, 500 000 $m^3$ of sawnwood is predicted to be produced from 1 000 000 $m^3$ of roundwood using the 2.0 input/output factor, which is represented in the reciprocal format of 50%. One might assume that 500 000 $m^3$ of sawnwood would have the same solid wood equivalent volume but this is not the case because wood products (including sawnwood) are often more or less dense than the parent wood used as the raw material. The 500 000 $m^3$ of dried sawnwood was 530 000 $m^3$ in the green state (prior to shrinkage), so a 1.06 factor is used to determine swe. Note that the total production of wood products in swe plus the losses (1 593 000 $m^3$) balances with the roundwood consumed (excluding bark volume and wood energy produced).

* The biomass power plant in this example is assumed to be operated entirely from bark volume, which is excluded from the balance.

**Source:** UNECE/FAO, 2010a.